NETFLIX NATIONS

T0252448

CRITICAL CULTURAL COMMUNICATION

General Editors: Jonathan Gray, Aswin Punathambekar, Adrienne Shaw

Founding Editors: Sarah Banet-Weiser and Kent A. Ono

Netflix Nations

The Geography of Digital Distribution

Ramon Lobato

NEW YORK UNIVERSITY PRESS

New York

NEW YORK UNIVERSITY PRESS

New York

www.nyupress.org

© 2019 by New York University

All rights reserved

References to Internet websites (URLs) were accurate at the time of writing. Neither the author nor New York University Press is responsible for URLs that may have expired or changed since the manuscript was prepared.

Library of Congress Cataloging-in-Publication Data

Names: Lobato, Ramon, author.
Title: Netflix nations : the geography of digital distribution / Ramon Lobato.
Description: New York : New York University Press, [2018] | Series: Critical cultural communication | Includes bibliographical references and index.
Identifiers: LCCN 2018021508| ISBN 9781479841516 (cl : alk. paper) | ISBN 9781479804948 (pb : alk. paper)
Subjects: LCSH: Netflix (Firm) | Video-on-demand. | Streaming video. | Television broadcasting. | International broadcasting.
Classification: LCC HD9697.V544 N48495 2012 | DDC 384.55/502854678—dc23
LC record available at https://lccn.loc.gov/2018021508

New York University Press books are printed on acid-free paper, and their binding materials are chosen for strength and durability. We strive to use environmentally responsible suppliers and materials to the greatest extent possible in publishing our books.

Manufactured in the United States of America

10 9 8 7 6 5 4 3 2 1

Also available as an ebook

CONTENTS

PREFACE

International television flows ("travelling narratives" in my re-definition) can be seen in a new light . . . as *flows of symbolic mobile and mobilizing resources* that have the potential to widen the range of our imaginary geography, multiply our symbolic life-worlds, familiarize ourselves with "the other" and "the distant" and construct "a sense of imagined places": in short, to travel the world and encounter "otherness" under the protection of the mediated experience.

—Milly Buonnano, *The Age of Television: Experiences and Theories*, 108–109 (emphasis in original)

As Milly Buonnano reminds us, watching television always involves some kind of imagined interaction with faraway places, situations, and symbols, in a way that recalls the word's etymological origins ("tele-vision": seeing at a distance). This idea of television, as an inherently international medium characterized by a particular way of ordering space, is at the heart of this book. In what follows, I revisit some long-standing debates in television and global media studies to see how they can help us understand the rapid transformations that are taking place as television morphs unevenly into an online medium.

Think of this book as an internet-era update to the rich literature on international television flows—a book for cord-cutting students and scholars who are interested in this longer history. Our central case study is Netflix, the world's largest subscription video-on-demand service. We will examine how Netflix morphed from a national media company to an international one between 2010 and 2016 and consider what this case means for existing debates about global television on the one hand and digital distribution on the other.

Both topics are of personal interest to me. I grew up in Melbourne during the 1980s and 1990s, watching a lot of television. Australia is a country where the local is always experienced alongside and through the imported—mostly American and British popular culture, but also some European and Asian content. Television has always been an international medium here. Local sitcoms share the schedule with U.S. network series, Hollywood movies, BBC tele-movies, and (on our public-service channels) the occasional Japanese cooking show or German police drama.

Australian television was broadcast-only until the 1990s. Even now, most Australians do not have cable or satellite subscriptions (though they are prodigious users of digital services, including pirate networks). In the late 1990s, I traveled overseas for the first time and observed the many ways that people watch television in other countries. Staying with relatives in Spain, I watched episodes of *Ally McBeal* in a dual-language track (a fascinating novelty to me but unremarkable to my Spanish cousins). In Morocco, I had my first taste of satellite TV, watching *The Simpsons* in French, courtesy of my friend's rooftop satellite dish. These were instructive experiences for someone used to Australia's five-channel TV environment—an environment that now seemed rudimentary compared to the denser distribution landscapes available overseas.

All this got me thinking about the relationship between television distribution, space, and culture. These issues would stay with me as a background fascination for many years while I researched in other areas, including film distribution and piracy, before returning to the topic of television when I started teaching classes on global media. Conversations with my students—who had many fascinating views to share on evergreen topics such as local content and cultural identity, and who themselves watch TV in the most diverse and interesting ways—prompted me to think once more about the difference that space makes to television culture. This book is the outcome of those many class discussions, filtered through the debates about streaming that were exploding around us when Netflix came to Australia, belatedly, in 2015.

The impact of Netflix in Australia was immediate and profound. Within a year, Netflix had attracted as many subscribers as our pay-TV service, Foxtel, which has been operating for more than 20 years (Roy Morgan Research 2016). Roughly one in three Australians now have access to Netflix at home (Roy Morgan Research 2017). Even before the service had officially arrived, thousands of Australians were using VPNs (virtual private networks) to illicitly access the U.S. Netflix library. Overall, the demand for Netflix in Australia has been remarkable.

This is certainly not the case everywhere—indeed, this book is substantially about Netflix's failure rather than its success in various markets around the world. Nonetheless, I found the experience of seeing a national television market so rapidly and thoroughly transformed by a foreign entrant affecting on many levels. This encouraged me to think more about the political, economic, and cultural impacts of streaming services. It also made me curious about connections between these services and the longer history of transnational television distribution via satellites. These were the fascinations that stayed with me as I wrote this book—a work of theory and analysis, based on a case study of a single platform, that explores the conceptual implications of internet distribution for global television studies.

There are many aspects of the topic that I have not been able to cover in depth here—including Netflix's original production strategy, which really deserves its own book. I am also aware that much of what I have discussed may have shifted by the time this book appears in print. Given these constraints of space and time, the book does not claim to offer a comprehensive account of Netflix—and it is certainly not an insider account (c.f. Keating 2012). Instead, it offers a selective analysis of what I see as the most important issues raised by Netflix's internationalization. The landscape will continue to change around us, but I hope these underlying issues will endure as central concerns for critical media scholarship.

NETFLIX NATIONS

Introduction

Every year in January, thousands of technology executives, geeks, and journalists make their annual pilgrimage to Las Vegas for the Consumer Electronics Show (CES). This massive four-day trade fair, one of the largest in the world, is where major brands such as Samsung and Sony show off their latest smart TVs, wearables, and other gadgets. In 2016, CES attracted over 170,000 people, including representatives from more than 3,000 technology companies. One of the keynote speakers was the CEO and cofounder of Netflix, Reed Hastings.

Hastings—joined on stage by Chief Content Officer Ted Sarandos and a number of Netflix stars—delivered the promotional spiel for Netflix's latest user-experience improvements and its new slate of original programming, playing clips from *The Crown* and *The Get Down*. At the end of the 48-minute showcase, Hastings made a surprise announcement: Netflix, long known for its patchy availability from country to country, was now a fully global television service, unblocked and accessible (almost) everywhere. "Today," said Hastings, "I am delighted to announce that while we have been here on stage here at CES we switched Netflix on in Azerbaijan, in Vietnam, in India,

Figures 1.1 and 1.2. Reed Hastings on stage at the Consumer Electronics Show, January 6, 2016, at The Venetian, Las Vegas. Photos by Ethan Miller/Getty Images.

in Nigeria, in Poland, in Russia, in Saudi Arabia, in Singapore, in South Korea, in Turkey, in Indonesia, and in 130 new countries. . . . Today, right now, you are witnessing the birth of a global TV network."[1] Reading from his teleprompter against a backdrop of world maps and national flags, Hastings went on to describe how this "incredible event" would make the Netflix experience available in the farthest reaches of the globe—everywhere, that is, except China ("where we hope to also be in the future"), North Korea, Syria, and Crimea (the latter three being countries where Netflix could not legally do business because of U.S. trade sanctions). "Whether you are in Sydney or St. Petersburg, Singapore or Seoul, Santiago or Saskatoon, you now can be part of the internet TV revolution," he promised. "No more waiting. No more watching on a schedule that's not your own. No more frustration. Just Netflix."

This announcement signaled a turning point for Netflix. Since the company first unveiled a streaming service for its U.S. customers in 2007, there had been speculation about when the company would offer this service to subscribers outside the United States. The rumors were confirmed when Netflix began its international rollout, first to Canada in 2010, then to Latin America in 2011, to parts of Europe in 2012 and 2013, and to Australia, New Zealand, and Japan in 2015. During this period, Netflix evolved from a national service (supplying American screen content to American audiences) into a globally focused business with greater ambitions. With the culmination of this process announced at CES, Netflix had become a global media company—available almost everywhere, with a potential foothold in almost all the major national markets.

Much of the world has embraced Netflix, and series such as *Stranger Things* and *Narcos* have amassed cult followings in many countries. Yet Netflix's metamorphosis into a global media provider has not been trouble-free. Shortly after Hastings's announcement, newspapers in a number of countries started reporting angry reactions to the Netflix global switch-on. In Kenya, the chairman of the Film Classification Board threatened to block Netflix on the grounds of its "shockingly explicit eroticism," arguing that "we cannot afford to be [a] passive recipient of foreign content that could corrupt the moral values of our children and compromise our national security" (Aglionby and Garrahan 2016). In Indonesia, access to Netflix was blocked by the state-owned telecommunications company (telco) Telekom Indonesia because of "a permit issue" and the "unfiltered" nature of its content (Gunawan 2016). In Europe, where there is a long history of cultural policy designed to keep Hollywood's power in check, regulators planned a minimum European content quota for foreign streaming platforms. Meanwhile, Australians fretted that the arrival of Netflix would "break" the internet as streamers hogged the bandwidth on the country's creaking internet infrastructure.

Stories such as these give us a sense of the diverse ways that countries have responded to the entry of Netflix into their media markets. They also show how Netflix's rise has revived some deep-seated tensions in international media policy. These tensions stem from differing views on the part of regulators, media companies, and audiences about the nature of video and its proper modes of distribution. They also involve disagreement about *where* video services should operate, *which* territories and markets they should be able to access, and *whose* rules they should obey.

This book takes the international rollout of Netflix as the starting point for a wider investigation into the global geography of online television distribution. By geography, I mean the spatial patterns and logics that shape where and how internet-distributed television circulates and also where and how it does *not* circulate. The book is organized around two central questions: How are streaming services changing the spatial dynamics of global television distribution, and what theories and concepts do scholars need to make sense of these changes? In answering these questions—one descriptive and the other speculative—this book will move across several subfields of media and communications research, including global television studies, media industry studies, and media geography. Along the way, we also delve into the history of earlier systems for transnational television distribution (especially satellite) and consider how they raised similar questions in the past.

Understanding Internet-Distributed Television

The rise of what Amanda Lotz describes in her book *Portals* (2017a) as "internet-distributed television"—professionally produced content circulated and consumed through websites, online services, platforms, and apps, rather than through broadcast, cable, or satellite systems—is an excellent opportunity to bring together two previously disconnected strands of television scholarship. The first of these is the rich body of literature about global and transnational television, which focuses on the connections (and irreconcilable differences) between institutions, practices, textual forms, and viewing cultures around the world (Barker 1997; Parks and Kumar 2003; Straubhaar 2007; Chalaby 2005, 2009; Buonnano 2007). The second is the literature on television's digital transformations, which explores the recent history of television technologies and their cross-pollination with other media and internet technologies (Spigel and Olsson 2004; Bennett and Strange 2011; Murphy 2011; Lotz 2014, 2017a). The arrival of internet-distributed television requires a rethinking of the potential connection between these two fields and their underlying categories: space and technology.

It is not merely that the future of television looks rather different in different places (Turner and Tay 2009, 8), although this is certainly true. Rather, internet distribution of television content changes the fundamental logics through which television travels, introducing new mobilities and immobilities into the system, adding another layer to the existing palimpsest of broadcast, cable, and satellite distribution. Internet television does not replace legacy television in a straightforward way; instead, it adds new complexity to the existing geography of distribution.

The arrival of mature internet-distributed television services such as Netflix is significant in global media debates. Until direct-broadcast satellite systems arose in the 1980s, television signals were mostly contained within national boundaries.[2] Even though programming was traded internationally, television distribution did not yet have a strongly transnational dimension. Recall Raymond Williams's classic anecdote in *Television: Technology and Cultural Form* (1974) about sitting in a Miami hotel room and watching American broadcast television, with its unfamiliar and disorienting "flow," for the first time. While Williams was familiar with American television as an imported medium, its actual broadcast distribution was something he could only experience by traveling to the United States.

One can only guess what Williams would make of today's television landscape, with its dizzying array of platforms and on-demand content. Today, one no longer needs to travel overseas to access international television, for a great deal of it is easily accessible online (under certain circumstances, and with many gaps and restrictions, which we will consider later). Similarly, the circulation of content is no longer determined by broadcast and satellite signal reach. The advent of internet-distributed television services means that it is now significantly easier for audiences in many parts of the world to view content from overseas—and even in some cases to access live channels—through browsers, apps, and set-top boxes.

This online proliferation of content is one consequence of television's digital transformation. The internet has become a distribution channel and archive for a diverse range of content, scattered unevenly across hundreds of platforms and portals. The digital mobility of content raises questions for scholars and students of media distribution, and also requires a rethinking of some of the assumptions that lie at the heart of television studies, because television content now circulates through the same infrastructure as other media, including ebooks, music, short videos, feature films, and podcasts. This has a number of significant conceptual implications for television studies that will be examined throughout this book.

Internet-Distributed Television as an Ecology

The first step in our analysis is to disaggregate the ecology of services, platforms, set-top boxes, and apps that constitute the field of internet-distributed television. Internet distribution of television content is not a unitary phenomenon; it involves a wide array of different services, institutions, and practices. Consider the way many viewers in broadband-enabled areas, especially younger audiences, watch TV: they use Google to search across sites for relevant free video streams, moving between the bits and pieces of content scattered around free video sites; they use third-party apps that filter and suggest particular programs; they follow recommendations from friends' posts on social media; they have active and lapsed subscriptions to video portals, some of which may be shared with friends and family; and they may also purchase individual episodes or season passes on their laptops and phones. In addition, some may use BitTorrent and illegal streaming sites, or share downloaded episodes and full seasons via USB sticks, cloud storage, and Bluetooth transfers. Depending on where they live and how tech-savvy they are, they may also use a VPN (virtual private network) or a proxy service to access offshore media or get around government restrictions on digital media services.

A point that is not new but bears repeating is that an increasing proportion of the global audience now understands television as an online service dispersed across an ecology of websites, portals, and apps, as well as a broadcast and cable/satellite-distributed medium. Key elements of this distribution ecology include

- *online TV portals*, such as BBC iPlayer (United Kingdom), ABC iView (Australia), NRK TV (Norway), and

Toggle (Singapore), which are provided by major broadcast networks and cable/satellite providers through websites and apps and typically include some combination of new-release content, library content, and live channel feeds;

- *subscription video-on-demand (SVOD) services*, such as Netflix, Amazon Prime Video, Hulu, HBO Now, Hayu, and CBS All Access, which offer a curated library of content for a monthly subscription fee;
- *transactional video-on-demand (TVOD) services*, such as iTunes, Google Play Store, Microsoft Films & TV, and Chili, which offer sell-through content at different price points for purchase and/or rental;
- *hybrid TVOD/SVOD/free portals*, such as YouTube, Youku, and Tencent Video, which offer free user-uploaded and professional content plus an additional tier of premium content available through subscription or direct purchase;
- *video-sharing platforms*, such as Daily Motion, which offer a range of free, ad-supported amateur and professional content, often informally uploaded;[3]
- *informal on-demand and download services*, including BitTorrent, Popcorn Time, file-hosting sites (cyberlockers), and illegal streaming sites;
- *unlicensed live, linear channel feeds*, delivered through pirate websites, illegal TV streaming boxes, and live streaming services such as Periscope; and
- *recommender and aggregator apps*, such as JustWatch, that advise what content is available across the various services.

This ecology is interconnected and highly dynamic, and therefore difficult to measure. To give a sense of scale, the European Audiovisual Observatory's MAVISE database of online video services currently lists 546 free streaming services, 448 transactional services, 367 subscription services (including adult sites), and 28 video-sharing platforms.[4] There is a lot of leakage between these categories. For example, catch-up services are becoming more and more SVOD-like, adding recommender systems and personal logins, while SVOD services are becoming more and more like conventional networks by producing their own exclusive content. Meanwhile, YouTube, Youku, and other hybrid sites tend to absorb innovations from many directions, combining advertising-funded free content, original content, live streams, user uploads, and pirated material in the one platform. To make things more complicated, there are also a wide range of gaming consoles, set-top boxes, dongles, and media players (Apple TV, Playstation, Amazon Fire stick, generic Android streamer boxes, Kodi boxes) that aggregate content from various sources, further blurring the line between distributors, aggregators, channels, and hardware providers.

Recent scholarship in media and television studies draws our attention to different parts of this ecology for different analytical purposes. For example, Stuart Cunningham and David Craig (2019) and Aymar Jean Christian (2017) emphasize the centrality of open platforms (especially YouTube) and networked sharing practices to contemporary television industries, thus advancing an expanded definition of internet television that includes social media platforms. Lotz's (2017a) category of internet-distributed television is defined more narrowly to refer to portals for professionally produced content ("the crucial intermediary services that collect, curate, and distribute television programming via internet distribution," such as CBS All Access, Netflix, and HBO

Now). Catherine Johnson uses a distinct term, "online television," to refer to a larger category of "closed and editorially managed" services that distribute "actively acquired/commissioned content" (Johnson 2017, 10)—a definition that would include public-service broadcaster portals as well as commercial SVODs and AVODs, but not open video platforms. These different ways of defining internet television are all instructive because they bring into focus particular parts of the ecology. This book focuses specifically on SVOD, but it does so with the understanding that SVOD represents only one line of development within a wider ecology.

The present instability within television distribution is remarkable, although historical precedents do exist. Recall that broadcast television evolved as a hybrid medium combining prerecorded material, live programming, movies, short-form programming, and advertisements. Early television was an empty container into which existing art forms and business models could be poured. The internet is now doing something similar for television, absorbing its existing textual forms and associated business models and putting them together in new combinations. Present distinctions between some of these categories may soon be rendered obsolete, a question addressed further in Chapters 1 and 2.

While I am interested in these historical questions, my primary focus is on the international *geography* of online television distribution—the spatial patterns that determine the availability and unavailability of television content to audiences in different parts of the world. These patterns are highly complex and volatile. This book describes a number of different phenomena that may sometimes appear contradictory. For example, while internet distribution has created new forms of mobility for content and audiences, it has also served to *reduce* mobility in other cases (e.g., via geoblocking), leading to increased territorialization. The relationship between television and its intended "zone of consumption" (Pertierra and Turner 2013) is variously stretched, dissolved, and reinforced. I want to emphasize that this is not the same thing as saying television is now everywhere, that it has been spatially liberated or deterritorialized, that space does not matter, or that content now circulates in a totally friction-free manner. This is not the case at all. Television is still bounded and "located" in all kinds of ways, as we will see in later chapters. The more accurate claim would be that internet distribution has introduced a new degree of complexity into the existing ecology. As a result, we are starting to see different kinds of relationships emerge between television's fundamental spatial categories: territory, market, nation, and signal area.

Why Netflix?

Netflix is presently the major global subscription video-on-demand service. It is not, however, the first television service with global aspirations. Various transnational channels, including CNN, MTV, Al Jazeera, and CGTN,[5] came before it, along with quasiglobal digital platforms such as YouTube. In calling this book *Netflix Nations*, then, I am not suggesting that Netflix is popular in *every* nation; my point is that Netflix, as a multinational SVOD service that spans national borders and operates in a large number of countries simultaneously, represents a particular configuration of global television that requires study and theorization.

I am also interested in Netflix for a different reason—because it draws our attention to unresolved questions about media globalization more generally. Specifically, the Netflix case provides an opportunity to test, advance, and refine our conceptual models of "global television" and to rethink what this term means in a context of digital distribution. As a company that has internationalized very quickly, Netflix's story also tells us a lot about what happens when a digital service enters national markets, coming in over the top of existing institutions and regulations. Netflix, in other words, is a case study with larger relevance to ongoing debates in media studies about convergence, disruption, globalization, and cultural imperialism.

The early history of Netflix is well known. The company was founded in California in 1997 by a direct-sales executive (Marc Randolph) and a Stanford-educated entrepreneur (Reed Hastings). Its first offering was a mail-order DVD rental service that proved wildly popular with American movie-lovers. Netflix unveiled its own streaming service in 2007 and fought off archrival Blockbuster, which declared bankruptcy in 2010. Along the way, the company became famous for its data-driven strategy, lean management ethos, and Silicon Valley–style human resources policies, which combine new-economy working freedoms (including unlimited leave time) with extremely high performance expectations.[6]

Netflix's staged international rollout began with its most strategically important markets—Canada and Latin America. These were the low-hanging fruit for Netflix:

Canada is a high-income, majority English-language market adjacent to the United States, while most of Latin America has a single regional language (Spanish), a large middle class, decent cable infrastructure, and a strong familiarity with pay-TV. Having successfully trialed its SVOD model in these territories, Netflix then expanded into key markets in Western Europe (2013–2015), Japan (2015), and Australasia (2015). In most of these countries, Netflix established partnerships with local telcos and internet service providers (ISPs), licensed locally relevant content and prepared promotional activities to coincide with the launch. Finally, the global switch-on at CES in January 2016 took care of the other lower-value or otherwise difficult global markets that had not yet been covered.

Netflix is one of the few media brands of the internet era to penetrate so deeply into households and the broader popular consciousness that it has become a verb ("let's Netflix it," "Netflix and chill"). It is a quintessential Silicon Valley success story, bridging two of America's signature fascinations—home entertainment and e-commerce. But Netflix is still a media company that trades in the established film and television industries' intellectual property, and since 2013 it has also invested heavily in its own original content. Unlike YouTube and Facebook, Netflix distributes only professionally produced content rather than user-generated content.

More than half of Netflix's subscribers now live outside the United States, and that figure is increasing. To cater to local tastes, the company has licensed thousands of non-U.S. titles—from Indian Bollywood movies to Turkish dramas—for its increasingly diverse user base, and it has invested billions of dollars in producing its own content in 30 national markets. As Netflix continues to reach a wider international audience, the service becomes more geographically differentiated and localized. Titles appear and disappear, and catalogs shrink and expand, as the platform is accessed from different parts of the world. Languages, currencies, and library categories are all customized for each country.

Just as Netflix is changing, users are changing Netflix. The platform learns from its new global audiences, tracking tastes and viewing habits. As a result, different "cultures of Netflix," as Ira Wagman and Luca Barra (2016) describe them, are starting to emerge—different ways of using the platform, talking about it, and watching it.[7] These user data feed back into the company's strategic decisions about original programming, licensing, and marketing. Netflix, then, should not be seen as a static cultural object or one that is consistent from market to market. It is constantly evolving, acquiring new layers of use and association.

This book is not a corporate history of Netflix, nor is it an insider account. Instead, it studies the debates and discourses *around* Netflix: how the service has been received by audiences, industry, and regulators in various countries. Since 2013, I have been closely following Netflix's rollout, drawing on a range of public sources, including trade papers, technical documents, press releases, corporate filings, promotional videos and texts, online user discourse, government and third-party policy documents, and various other sources to piece together the story. I have also been fortunate to work with a number of talented, multilingual research assistants—Wilfred Wang, Ishita Tiwary, Renee Wright and Thomas Baudinette—who wrote reports on key territories (China, India, Russia, and Japan), providing vital context for the study. *Netflix Nations*, then, is a study of Netflix from the outside: a study of impacts, discourses, and debates, grounded in a tradition of critical media research. It makes no claim to get inside the black box or the boardroom.

I have written this book with several kinds of readers in mind. For students and scholars of television, it is first and foremost a book that tells a critical story about the world's largest SVOD service and what its international rollout has meant for television distribution and media policy. At a conceptual level, the book is about the problem of media globalization and the rich history of intellectual debate around it. Finally, it is also a reflection on the state of television research in the internet age. It asks how scholars in this field might engage critically and productively with challenging new issues—such as localization and search technologies, and internet policy and regulation.

The book is divided into seven chapters. Chapter 1, "What Is Netflix?," provides a critical survey of current debates in television studies and internet studies as they relate to digital distribution. It also discusses the ontology of new television services, tracing connections to a range of different media forms. Chapter 2, "Transnational Television: From Broadcast to Broadband," explores how debates about multinational and transnational television services have evolved over the years. Placing Netflix in a longer history of transnational television services, including broadcast and satellite channels, it explores how familiar anxieties about national sovereignty are returning in a different guise through internet distribution. Chapter 3, "The Infrastructures of Streaming," takes an infrastructural approach to understanding Netflix. Here we examine some of the platform's underlying systems, including its Content Delivery Network (Open Connect), and related policy issues such as net neutrality. Chapter 4, "Making Global Markets," considers how Netflix has attempted to enter diverse national markets and adapt its offering to conform to local audience expectations. Case studies of Netflix's experience in three key Asian markets—India, China, and Japan—reveal the challenges of localization and market entry. Chapter 5, "Content, Catalogs, and Cultural Imperialism," focuses on cultural policy debates relating to Netflix catalogs, especially regarding local content, and examines how regulators in the European Union (EU) and Canada are attempting to develop local content policies for over-the-top services. Chapter 6, "The Proxy Wars," tells the story of how Netflix sought to manage VPN use and geoblocking circumvention by users during the early years of its internationalization. I also consider how Netflix's policies on copyright and piracy evolved over those years. The book concludes with some reflections on parallel models of evolution in television industries beyond SVOD, including recent developments in China, which reflect a different pattern of transformation.

As this structure suggests, my aim in this book is to use the *controversies* that have swirled around the Netflix service as a starting point for building a theory about the relationship between global television and internet distribution. In this way, the book develops a series of arguments and analyses that position Netflix within a longer trajectory of debate, reaching back through the history of transnational television. Each chapter begins with a particular analytical problem relating to global media, such as infrastructure, cultural imperialism, or localization; considers how this problem plays out in the case of Netflix;

and then finally asks what Netflix can add to our understanding of these concepts. Netflix, in this sense, becomes a resource—or perhaps a *platform*—for revisiting enduring critical debates in global media studies.

NETFLIX

New Releases

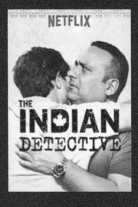

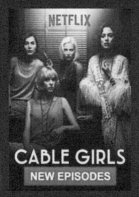

Watch It Again

TV Shows

1

What Is Netflix?

In the introduction to their book *YouTube: Online Video and Participatory Culture* (2009), Jean Burgess and Joshua Green make an important point about the challenges of studying emergent digital media. For Burgess and Green, one of the most interesting and difficult things about writing a book on YouTube was the fact that it was still evolving. Late in the last decade, YouTube had a chameleonic character: it was a "distribution platform that can make the products of commercial media widely popular" while at the same time being "a platform for user-created content where challenges to commercial popular culture might emerge" (Burgess and Green 2009, 6). Its creators, investors, and users—not to mention media academics—had yet to agree on what YouTube actually *was*, meaning that there was still much uncertainty over what the platform could be used for, how it should be regulated, and how it could be understood in relation to other media. Burgess and Green argue that

> because there is not yet a shared understanding of You-Tube's common culture, each scholarly approach to understanding how YouTube works must make different choices among these interpretations, in effect recreating

Figure 1.1. Netflix mobile interface, as of January 2018. Screenshot by the author.

it as a different object each time—at this early stage of research, each study of YouTube gives us a different understanding of what YouTube actually *is*. (6–7, emphasis in
original)

This basic ontological problem (what is a digital media
service, and how do we interpret and theorize it?) applies
to a range of phenomena that exist at the boundaries of
television, cinema, and digital media. Scholars studying
Netflix must therefore make certain choices about what
kind of service it is and what the appropriate frames of
analysis should be. These decisions work to re-create the
object anew each time by opening up or closing off lines
of comparison.

While Netflix is an established global brand with 20
years of history, there is still very little agreement about
what Netflix *is* or how it should be understood by the
public, scholars, or media regulators. Netflix—like many
disruptive media phenomena before it, including radio
and broadcast television—is a boundary object that exists between, and inevitably problematizes, the conceptual
categories used to think about media. This definitional
tension can be seen in the marketing slogans Netflix uses
to describe itself, which reflect evolution in both the company's distribution model and its discursive positioning in
relation to other media. Presently, Netflix defines itself as
a "global internet TV network," but in the past it has preferred terms such as "the world's largest online DVD rental
service" (2002), "the world's largest online movie rental service" (2009), and "the world's leading Internet subscription service for enjoying TV shows and movies" (2011).[1]
Others have referred to Netflix as "a renegade player in
the television game" (Farr 2016, 164), "a pioneer strad-

dling the intersection where Big Data and entertainment media intersect" (Leonard 2013), a "monster that's eating Hollywood" (Flint and Ramachandran 2017), and even "a company that's trying to take over the world" (FX CEO John Landgraf, cited in Lev-Ram 2016). Other possible responses to the question "what is Netflix?" might include

- a video platform,
- a distributor,
- a television network,
- a global media corporation,
- a technology company,
- a software system,
- a big-data business,
- a cultural gatekeeper,
- a lifestyle brand,
- a mode of spectatorship, or
- a ritual.

Clearly, Netflix means different things to different people. Part of the issue here is that there are a number of incompatible interpretive frames in use. Each frame brings with it a set of assumptions and invokes a particular history of industrial and technological evolution. As we move through these various descriptors, Netflix's location within industry sectors also seems to shift around—between the television, video, technology, internet, digital media, entertainment, and information industries. The conceptual frameworks we use to understand Netflix are important because they shape the kind of thinking we bring to the analysis. Consequently, these frameworks require some critical reflection.

This chapter traces out two different analytical perspectives that can be applied to Netflix and in so doing critically synthesizes two related fields of scholarly literature. The first of these can be found within television studies, in the form of research on TV's digital and postbroadcast transformations. The second comes from outside television studies, via new media theory, internet studies, and platform studies. As I will argue, it is helpful to move between and across these two ways of knowing so as to avoid the intellectual lock-in effects that result from following one line of thinking too closely. For example, if we study Netflix in terms of its similarities to and differences from television, we can miss its connections to other digital media. Similarly, focusing exclusively on the software dimension obscures Netflix's structural relationships with established screen industries. We need to be aware of the natural pull of particular ways of thinking and what they reveal and obscure when applied to different kinds of media objects.

Television Studies and the Future-of-TV Debate

Today, the academic field of television studies is in a state of flux as it undergoes another round of self-reflection. In recent years, a rich corpus of postconvergence research and theory has emerged to explore how digital technologies of various kinds have variously transformed, extended, and sustained existing television industries. This literature asks questions such as: What is television now? What might it become? Is what we used to call the "idiot box" dead, dormant, or as dominant as ever? In the age of televisual "expansion and overflow" (Gray 2009, 85, citing Brooker

2001), where do the boundaries around a medium, a distribution system, or an individual text lie?

Questions such as these have been carefully examined by scholars, including William Uricchio, Milly Buonanno, Chuck Tryon, Amanda Lotz, Lynn Spigel, and Graeme Turner, among others. A number of influential anthologies have appeared, including *Television after TV: Essays on a Medium in Transition* (Spigel and Olsson 2004), *Television Studies after TV* (Turner and Tay 2009), *Television as Digital Media* (Bennett and Strange 2011), and *After the Break* (de Valck and Teurlings 2013), as well as numerous monographs and trade books. Television studies journals, including *Television and New Media*, *Flow*, and *View*, have played host to vibrant debates about these issues. A wider body of technical and policy literature also exists, much of it authored by telecommunications experts; for example, Columbia University media economist Eli Noam has been writing about internet-distributed television since the 1990s, before it was of mainstream interest to media scholars.

Broadly, this literature maps an ongoing but uneven set of transitions in the history of television that are collectively working to transform it from a mass medium to a niche one through technological and institutional developments that "fragment the previously mass audience of television into a series of personalized choices" (Bennett 2011, 2). Kelsey (2010, 231) writes that, "We don't just watch TV, we send and receive it, gather and organize it on our personal touch screens, meanwhile interacting with sites to produce, wittingly or not, the consumer feedback that helps broadcasters determine a season's programming (if TV still even thinks in terms of seasons)." Tryon (2013, 14) argues that "contemporary media platforms actively solicit an individualized, fragmented, and empowered media consumer, one who has greater control over when, where, and how she watches movies and television shows," warning that "this offer of liberation from the viewing schedule is often accompanied by increased surveillance." In response to these shifts, alternative periodizations of television technology are also emerging. Some experts now refer to TVI (broadcast only), TVII (cable era), and TVIII (digital distribution), terms that draw our attention to the successive waves of transformation that have swept through television technology and the television industry (Todreas 1999; Pearson 2011; Johnson 2007).

The work of U.S. television scholar Amanda Lotz offers a richly textured account of these transformations. Across a number of books—especially the second edition of *The Television Will Be Revolutionized* (2014), *Portals: A Treatise on Internet-Distributed Television* (2017a), and *We Now Disrupt This Broadcast* (2017b)—Lotz provides a forensic examination of the changes in the underlying economic models of television when it moves online, and how these models shape programming, production, and circulation. Lotz begins by explaining how the fundamental logic of television has been predicated on linearity: "Almost all the conventions of television—a flow of content, program length, expectations of weekly episodes—derive from practices developed to cope with the necessity of the linear schedule" (Lotz 2017a, 15). In contrast, the on-demand character of internet-distributed television, and its precedents in earlier on-demand services (such as pay-per-view movies delivered by cable), presents a different mode of distribution that has more in common with the record store, bookstore, or library. In this way, internet-distributed television "allow[s] behaviors that were peripheral in an age of analog, physical media such as time shifting, self-curation, and à la carte access to become central and industrialized practices" (17).

Lotz sees Netflix as a central part of this story, not only because the company "disrupted the long acculturated sense that television content should be viewed on a television set" (Lotz 2014, 71) but also because it introduced new kinds of filtering, aggregation, and recommendation systems that have since become widespread. She points to the Netflix Queue (now called a List) as a key site through which users negotiated the shift to nonlinear television, noting that "the queuing that Netflix introduced provided its subscribers with a different paradigm for thinking about and organizing viewing behavior, and one that substantially challenges the long dominant, linear, 'what's on' proposition" (74). In other words, Lotz regards the online distribution of content as highly significant because it marks a transformation in the underlying structure and business models of television by freeing content from a linear schedule and by introducing new pricing models (including all-you-can-stream subscription packages) and audience expectations about the content, novelty, and value of TV services. As she writes, "The affordance of internet protocol technologies to deliver personally-selected content from an industrially curated library is the central difference introduced by this new distribution mechanism" (Lotz 2017a, 4).

Within the various contributions to the future-of-TV debate, we can see different degrees of emphasis on change as opposed to continuity. Lotz foregrounds the transformative dimensions of internet distribution in her work, while other scholars focus on the continuities. In this

second category, we often find the work of media historians, who are—by training and temperament—ambivalent
about diagnoses of radical change. William Uricchio, for
example, stresses that notions of personal TV and interactive TV go back much further than the internet era and
can be traced right through the history of the medium,
with precursor concepts to be found throughout the twentieth century:

> Television offers a striking case where both the techno-
> logical platform and its deployment protocols have shifted
> radically and more or less continually since the late 19th
> Century. We've seen the project of the televisual ally it-
> self with platforms such as the telephone, radio, film, and
> networked computer; and we've seen its protocols include
> person-to-person communication, entertainment and
> news, surveillance, telepresence and so on (not to mention
> legal and regulatory rule sets). (Uricchio, forthcoming, 11)

Uricchio reminds us that if we wish to understand the
future of television we do not have to start with the internet. Instead, we can look back to early video game technologies, the introduction of cable and satellite systems, the
VCR and TiVo, and even the remote control—all of which
have contributed in different ways to television's personalized, postbroadcast present by variously expanding the
range of content available, increasing viewer control over
the flow of images, and introducing elements of interactivity (Wasser 2002; Boddy 2004; Uricchio 2004; Thomas
2008). Following Uricchio, we can look back even further,
to a range of visionary early twentieth-century experimen-
tal television technologies that prefigured "what in today's
terms might be understood as Skype, surveillance video,

large screen public display, and domestic news and entertainment" (Uricchio 2004, 7–8). This is why many scholars who use terms such as postbroadcast and postnetwork are careful to emphasize that they signify not epochal change (from X to Y) but rather the sedimented layering of different technologies, systems, institutions, and viewing cultures, such that cable, satellite, internet, and mobile technologies coexist with and are structurally integrated into broadcast television (Turner and Tay 2009; Parks 2004; Lotz 2014).

A second lesson from this literature is that we should not write off the institutional power of television just yet. Toby Miller lucidly argues that television as an industry sector is far from dead—and anyone who claims otherwise is likely to be proven wrong by history. Miller is highly critical of the death-of-TV discourse and mocks the assumption that "the grand organizer of daily life over half a century has lost its pride of place in the physical layout of the home and the daily order of drama and data" (Miller 2010, 11). Instead, he emphasizes the industrial continuities (especially in production and advertising) that persist into the internet age. Miller offers a series of counterarguments in response, noting that a lot of internet media is basically television; that television institutions are still structurally central to digital media markets; that broadcast television is still strong and important globally; that there are more TV stations opening up worldwide than ever before, especially in emerging economies; and that audience ratings suggest we are actually watching more television content than ever before (it is just distributed differently).

This is indicative of one response to the future-of-TV debates, which is to affirm the centrality and vitality of

television institutions in the face of their digital dethrone-
ment. As Tim Wu (2015) reminds us, "Overestimating
change in the television industry is a rookie mistake."
A different formulation of the argument can be found in
media business commentator Michael Wolff's book *Television Is the New Television* (2015). Setting out to destroy
what he sees as the Silicon Valley myth of television's disruption at the hands of the digital, Wolff argues that the
recent history of media is better understood the other
way around—that television has ultimately tamed and
absorbed digital media. For Wolff, Netflix is a classic example of this reverse engineering of the digital. The service
is much more television-like than internet-like, Wolff argues, because it shuns many of the interactive affordances
of internet media in favor of established narrative structures, aesthetics, and experiences. In Wolff's account, as
Netflix morphed from a DVD rental company to a digital
studio, it actually moved closer to television by "bringing television programming and values and behavior—
like passive watching—to heretofore interactive and
computing-related screens" (Wolff 2015, 91). He adds:

> Other than being delivered via IP, Netflix had almost
> nothing to do with the conventions of digital media—in
> a sense it rejected them. It is not user generated, it is not
> social, it is not bite size, it is not free. It is in every way, ex-
> cept for its route into people's homes—and the differences
> here would soon get blurry—the same as television. It was
> old-fashioned, passive, narrative entertainment. (93–94)

In this argument, we can see a variation on the future-
of-TV arguments: the idea that television has already
shaped the future of digital media in its own image and

will continue to exert influence on audience expectations and industrial norms. In Wolff's view, internet television services—while introducing all manner of new innovations—are likely to succeed only to the extent that they offer television-like experiences and a corresponding value proposition. In this sense, television is fated to live on both as a resilient industry sector and as an experiential gold standard that will shape audience expectations regarding content, no matter what transformations take place at the point of distribution.

For industry analysts, there is much at stake in predictions of industry change. For media scholars, the key issues are somewhat different and also require consideration of the agency of particular arguments about what television was, is, and might become. If we follow Uricchio in thinking that television has never been ontologically or technologically stable but can only be *stabilized* to a greater or lesser degree, then the critical question for media scholarship becomes identifying the ways in which particular discourses of change and continuity operate to lend a "conceptual coherence" to a medium or technology at particular points in time (Uricchio, forthcoming, 7–8; Uricchio 2004). In other words, while we cannot predict the future of television in the internet age, we can try to understand how particular ways of thinking about that future might help to shape the way such a future—or range of futures—will play out.

For this reason, it is necessary for certain branches of media scholarship to become more self-reflexive about their own investment in the object of television as a discrete medium and in television studies as a discrete field of inquiry. As Lotz writes in *The Television Will Be Revolutionized*, "In many ways, HBO and Netflix are more alike because they are non-advertiser-supported subscription services than different because one comes in through cable and the other over broadband—a distinction I suspect will be technologically nebulous the next time I revisit this book" (Lotz 2014, xii). From the point of view of media studies, this raises questions about whether a platform such as Netflix *should* be studied as television and what is gained or lost in doing so. After all, most users of streaming services are still likely to think of professionally produced scripted content as television content, but they do not always watch it *on* the TV and perhaps do not care about whether it comes over the top, via cable, or over the airwaves. Nor may they be concerned about whether the analytical integrity of television-as-industry or television-as-medium has been compromised.

In a much-cited essay published more than a decade ago, Lynn Spigel asked, "What is to be gained from studying TV under the rubric of new media?" (Spigel 2005, 84). This question is still important, and largely unresolved, because it prompts us to think about what is revealed and obscured as one moves between ontological frameworks. One of the questions we need to think about is not whether the future of television is going to look more like the internet or more like cable but rather whether emergent media forms should be understood in terms of their similarities to past media or through entirely new paradigms. The trick may be to build productively on past knowledge without letting existing frames of reference overdetermine objects of analysis. Academic disciplines are slow; they rely on the incremental accumulation of knowledge. In the case of television studies, it is neither useful nor desirable to throw out all this knowledge and deep thinking behind it every time a new distribution technology appears on the horizon (as has already happened with video, Tivo, mobile devices, and so on). But, at the same time, there are some risks in trying to assimilate a wide array of convergent and new phenomena into an existing paradigm, just as there are risks in taking the reflex position that we have seen it all before and that it is all still television. Even though we can trace many paths between past and present, we also need to acknowledge the differences and find ways to come to terms with them analytically. This need is especially acute when the everyday terminology may remain unchanged ("watching TV") but might refer to quite a different set of practices that are ontologically distinct from what that terminology referred to in the past.

In grappling with the conceptual problem of internet television, then, we need to be alert to diverse and sometimes contradictory effects. On the one hand, it is quite possible that nonlinear internet distribution will come to function primarily as simply another distribution channel for existing content or new content that still looks and feels like TV as we know it. Seen from this perspective, internet distribution can reasonably be understood as something that is easily assimilated into existing business models. But there are longer-range effects at work here, too, and not all of them can be predicted in advance. Over time, the nonlinear affordance of internet distribution is likely to lead to further specialization and expansion in content production, such that new texts may increasingly be designed for the experiential specificities of internet rather than broadcast or cable distribution. We can already start to see this with the kinds of quality dramas made explicitly for binge viewing, and in the proliferation of short-form web comedies that would not fit well into a traditional schedule, not to mention the vast pool of amateur content on YouTube. This suggests that changes in distribution can have longer-term effects in other areas of the system, including production and reception. While we may still watch TV in familiar ways, in familiar spaces and formats, transformations are taking place that slowly recalibrate the whole system.

The question "is it still TV?" is problematic precisely because its framing invites a reductive "yes" or "no" answer that works to solidify a category (television) that may instead be better deconstructed, or at least reformulated. Cunningham and Silver (2013) argue that instead of asking whether new media has changed old media, and thus lapsing into familiar binaries of technological crisis versus continuity, we should focus instead on how to account for the *rate* of change, and the particular combinations of change and stasis that exist at any one time in the history of a medium. They reject both the "everything has changed" and "nothing has changed" positions as inadequate responses to the question of media industry transformation. Following this lead, we could also ask what other intellectual resources are available to us for thinking about the relationship between, rather than merely the "impact of," internet distribution vis-à-vis television.

An excellent model is provided by Thomas Elsaesser and Malte Hagener, who have worked through this problem in a different context. In their chapter "Digital Cinema and Film Theory—The Body Digital," they extrapolate Lev Manovich's idea of the *inside-out* to advance the argument that digitization can create simultaneous stasis and change, leading to their apparently paradoxical conclusion that, "With digital cinema everything remains the same and everything has changed" (Elsaesser and Hagener 2015, 202). What Elsaesser and Hagener mean by this is that there has been great change within the boundaries of an existing category such that the referent of the category itself is transformed and we are no longer talking about the same thing we thought we were talking about. Hence it is not so much a matter of tracing lines of continuity and change around a fixed axis but rather grappling with the "inside-out" ontological transformation of a medium.

Consider how Elsaesser and Hagener work through this paradox in relation to cinema. Their account insists that the social experience of cinema-going remains popular, durable, and powerful ("stars and genres are still the bait, concessions and merchandise provide additional or even core revenue for the exhibitor, and the audience is still offered a social experience along with a consumerist fantasy," 202). However, they also claim that the textual form of digitally shot cinema has been reorganized through digital production, such that the relationship between image and representation is now completely recast. Digital cinema now *produces the effect* of cinematic representation as just one of its potential applications. Hence there is not only a combination of stasis and change but also a series of internal changes that produce the same external appearance. This is change from the "inside out," such as when a parasite takes over its host, "leaving outer appearances intact but, in the meantime, hollowing out the foundations—technological as well as ontological—on which a certain medium or mode of representation was based" (204–205).

This is a compelling theory of technological change, a reminder that change and stasis not only coexist but can also envelop each other. Elsaesser and Hagener are referring to production techniques in the main. However, there is some parallel to distribution. The inside-out transformation of internet television allows the TV experience (the reception technology, domestic space, textual formats, and so on) to remain consistent with established norms while unfurling a substantive change on the inside—specifically, the inherent *nonlinearity* and *interactivity* of the digital video platform. Viewers pick and choose individual items from a database rather than watch what is "on" at any given time. There is no scheduled flow of programming (even though much of the licensed content offered on SVOD services was produced for such a schedule); there are only individual pieces of content within a database of offerings that can be consumed in any order, at any time, and that will often continue to play automatically thanks to the Netflix Post-Play feature (which automatically cues the next episode). Depending on how we evaluate such structural changes in distribution, this aspect of internet television may indeed embody the same inside-out quality that Elsaesser and Hagener identify in digital cinema production.

As we can see from these various arguments, there are benefits and risks to seeing Netflix through the lens of television. Such a perspective opens our eyes to important continuities in the experience, production norms, and domestic context of moving-image entertainment, but it can also produce some analytical traps. This is why it is helpful to take a *both/and* approach, so we can approach our object from multiple perspectives simultaneously. As we have seen, Netflix may still feel like TV to viewers, and it relies on this familiar pleasure for its success, but its distributional logic is markedly different—technologically, economically, and structurally. It is too early to tell, of course, but we should at least entertain the possibility that the affordance of internet protocol distribution may well prove to be the parasite inside the host—the agent of change that ends up quietly overtaking the organism from the inside out—while still retaining its outward features.

Digital Media Studies and the Platform Perspective

Let us consider a second analytical approach and what it might bring to an understanding of Netflix. This second approach would consider Netflix as a *digital media service*—a computational, software-based system that can produce a television-like experience as just one of its potential applications. Following this line of thought—which in fact aligns with the historical origins of the company and the way it presents itself to investors and regulators, if not to users—we can start to see how Netflix fits in with a quite different set of debates that have been playing out in fields such as new media studies, internet studies, and platform studies. In this section, I explore some of the arguments relevant to Netflix that have emerged from these debates. This will push our analysis of Netflix in a direction different from where television studies might take us.

This second way of thinking is less concerned with understanding Netflix *in relation to* television, cinema, or any other form of screen media, however one defines it. In contrast, it sees Netflix as a complex sociotechnical software system. It is more interested in looking sideways to other digital media, rather than backward to television, to assess similarities and differences. There is, then, a fundamental difference between a television studies approach and a digital media approach. The former is inherently historicizing; it sees its object in relation to a particular media technology (television) and its evolution. In contrast, the latter implicitly frames its object as a set of computational technologies tied together into a common user interface while also understanding each digital media service as a kind of communication system in its own right—with unique design, affordances, and limitations. This allows us to think about Netflix alongside a much wider range of *digital media*, including not only video platforms (You-Tube, Youku, Hulu) but also e-commerce and social media networks (Facebook, Twitter, Ebay, Amazon, Weibo) as well as other software artifacts, such as electronic program guides (EPGs), gaming consoles, or desktop operating systems.

The term "platform" requires some explanation. In new media and internet studies, platforms are commonly defined as large-scale online systems premised on user interaction and user-generated content—including Face-book, Twitter, Medium, Snapchat, YouTube, Flickr, Grindr, and others. Platform studies, as it has become known, is a field of critical, empirical, and theoretical research concerned with these new institutions of the internet age and the specific ways in which they have been able to harness user communication and labor. It seeks to understand how platforms mediate and organize our daily interactions, asking what this means for communication practices, economies, and identities. Of course, Netflix is not a platform in the same way as social media services like Facebook or Twitter are. Netflix is not open, social, or collaborative. One cannot upload content to Netflix or design software applications to run within it. In this sense, it is fundamentally different from video sites containing both user-uploaded and professionally managed content (YouTube, Youku, etc.). Unlike these sites, Netflix does not (at this stage) have an advertising business model; nor does it have the character of a multisided marketplace like Amazon or Ebay, which host a more complicated ecology of commercial activity. Netflix is closed, library-like, professional; a portal rather than a platform; a walled garden rather than an open marketplace. This said, we can still learn a lot *about* Netflix through platform studies perspectives.

Platform studies has evolved along two main lines. The first of these comes out of the work of Nick Montfort and Ian Bogost. In their book *Racing the Beam: The Atari Video Computer System* and related working papers—which are widely read in games studies, though less so in television studies—Montfort and Bogost outline a specific understanding of platforms and how they can be studied. They define a platform as the "hardware and software framework that supports other programs" (Bogost and Montfort 2009a, 1) and as "a computing system of any sort upon which further computing development can be done" (Bogost and Montfort 2009b, 2). They note that a "platform in its purest form is an abstraction, simply a standard or specification" (Bogost and Montfort 2009a, 1). Consequently, their vision of platform studies involves "investigating the relationships between the hardware and software design of standardized computing systems and the creative works produced on those platforms" (ibid.). Montfort and Bogost insist that researchers pay close attention to the materiality of the platform, including its design, construction, and even wiring, as well as to the platform's user-facing and symbolic dimensions. Their approach is better suited to gaming systems such as Atari and PlayStation than to online services like YouTube, Steam, or Netflix—though the material dimensions of the latter are also amenable to research and critique, as we will see in Chapter 3.

A second strand of thinking about platforms comes out of critical communications and internet research. The work of Tarleton Gillespie in particular draws our attention to the expanding range of everyday communication and consumption practices that take place within online platforms, especially social media networks. Gillespie defines platforms as "sites and services that host, organize, and circulate users' shared content" (Gillespie 2017, 254). His essay "The Politics of 'Platforms'" (Gillespie 2010) was an early critique of the way online services such as Facebook and YouTube strategically defined themselves as neutral intermediaries—as technology companies rather than media companies—thus obscuring their power as mediators of communication, identity, and politics.[2]

A key theme in Gillespie's work is the agency of the platform itself. Far from being neutral, platforms shape the communications, interactions, and consumption that they facilitate—through interface design, moderation policies, terms of service, algorithmic recommendation, and so on. Consider the Facebook "Like" button and how it subtly institutes a norm of extroverted positivity as the default practice for online communications—there is no "Dislike" or "Don't Care" button—while at the same time generating valuable commercial data for Facebook by turning "personal data into . . . public connections" (van Dijck 2013, 49; Gerlitz and Helmond 2013). We should not, then, make the mistake of seeing a platform as a "neutral" distributor of content, because the nature, design, and business model of the platform will always have an effect on what passes through it. Platforms, according to Gillespie,

> have precise (and shifting) technical affordances that constrain and guide practice—both in their own design and in their fit with a myriad of infrastructures, including their back-end data systems, the protocols of the Web,

and the dictates of mobile providers. They have rules and norms that bless some practices and are used to restrict others. They have myriad international, sometimes conflicting, legal obligations they must enforce. They have commercial aspirations and pressures that drive decisions about how they're marketed, how they're updated, and how they're positioned against their competitors. (Gillespie in Clark et al. 2014, 1447)

Following Gillespie's arguments, it is possible to see how Netflix—while certainly not a social media platform—exploits the same quality of discursive slipperiness as these other platforms. Netflix, like Facebook and YouTube, is presently engaged in a number of disputes with government agencies about how and whether it should censor its film and television content. In India, for example, Netflix claims that because it is an internet-delivered service rather than a broadcaster, it should not have to follow the obscenity policies that apply to Indian television stations (see Chapter 4). This is not all that far from Uber's insistence that because it is a technology platform it should not have to follow the licensing and tax laws that apply to taxi companies, or Facebook's insistence that it is not a media company and therefore should not have to fully regulate the communications taking place through its networks. In each case, a service's digital status is invoked to sidestep regulatory responsibilities.

Even though these three companies operate in very different markets (transport, communications/advertising, and scripted entertainment), they have a common operational logic that hinges on their status as a digital service that is (a) categorically dissimilar to the established incumbents they now compete with and (b) operating in global markets from a U.S. base, partially outside the jurisdictional reach of national governments. Following this logic, and notwithstanding the lines of historical evolution between Netflix and television traced in the previous section, one can also argue that these structural similarities with other digital services place Netflix within the platform economy as much as within the entertainment industries.

Another common characteristic of digital media platforms is a reliance on algorithmic recommendations. Along with Amazon and Pandora, Netflix has played a pivotal role in the development and popularization of recommendations generally, having invested heavily in this area since its years as a DVD rental service. The company famously ran an open engineering competition, the US$1 million Netflix Prize of 2006–2009, to improve its predictive powers by 10%. The fruits of these efforts have paid off in the form of its eerily accurate prediction engine, which seeks to, in Hastings's words, "get so good at suggestions that we're able to show you exactly the right film or TV show for your mood when you turn on Netflix" (*The Economist* 2017). On the Netflix home screen, algorithmic recommendations are used to autocurate selections of content geared around individual users' data profiles. Every video selection that appears on the home screen is the result of intricate calculations based on user-submitted data (movie ratings and viewing history), collaborative filtering (predictions based on other people's activities), and manual coding of films for all conceivable metadata points, from character types to endings.

This naturally puts Netflix squarely in the middle of debates about the datafication of culture, filter bubbles, and big-data politics (Pariser 2011; boyd and Crawford 2012; Beer 2013). Its recommendation system has been accused of everything from unjustified consumer surveillance to the demise of the mass audience and the end of serendipity. Film scholars in particular have voiced concern about the way personalization leads to filter bubbles. In an essay on Netflix's "mathematization of taste," Neta Alexander (2016, 94) warns that "the rise of predictive personalization might be good news for the study of artificial intelligence and machine learning, but it is bad news for anyone who wishes to encounter what Sontag calls 'great films.'" We should, however, bear in mind that algorithms can be programmed for diversity as well as for taste reproduction (Blakley 2016).

The debate about Netflix's effect on taste and consumption continues to rage, though it is not a primary focus of this book. For our purposes, let us instead focus on the design of the Netflix interface and how this mediates relations between television, cinema, and digital media. The Netflix interface changes regularly but at the time of writing is organized into categories that are curated automatically from a list of thousands of potential options, including popular genres (romantic comedies) as well as hyperspecific microgenres (fight-the-system documentaries) (Madrigal 2014). This smorgasbord of content is arranged into celluloid-like strips of color that slide off the right-hand side of the page, suggesting an infinite variety of choices. In this way, the viewer is positioned as the sovereign navigator-user of an endless archive of screen content. Such design choices are carefully constructed to create the appearance of textual abundance and conceal limitations in what is a finite Netflix catalog.

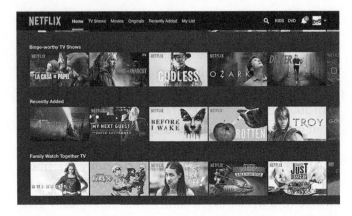

Figure 1.2. Netflix desktop interface, as of January 2018. The interface, designed in such a way as to conceal catalog limitations, suggests an endless bounty of content available to the user. Screenshot by the author.

Until 2015, the Netflix desktop interface had a light grey background. Video artwork was formatted in vertical, DVD-style boxes, so that the overall effect was reminiscent of a video store. Now, the background is dark—as in a movie theater—and the DVD covers have been rearranged into a horizontal format suggesting frames on a celluloid filmstrip. This site update seems designed not only to make the service as tablet-friendly as possible, hence the shift to the horizontal format, but also to discursively reposition the site within the pantheon of older media technologies by moving the idea of Netflix away from video-store and DVD culture— surely a fading memory for most of its users—and realigning the service with that most resilient medium, cinema. Interestingly, the iconography of television is nowhere to be found in Netflix's interface design, despite the abundance of TV series available through Netflix. There are no remote controls, advertisements, or schedules. Even though the idea of television is central to Netflix's commercial ambitions—recall Hastings's description of Netflix as "a new global Internet TV network"—the television experience does not seem to be central to how Netflix wishes its users to imagine streaming. Perhaps this is because of the degraded nature of the "idiot box," and Netflix's related desire to market itself as a premium service. In any case, it is one of the ironies of internet television that its referent medium, television, is being simultaneously reimagined, integrated, erased, and remediated through the emergence of streaming services.

Toward a Synthesis

This brings us back to Netflix's relationship to screen media. As we have seen, Netflix is a shape-shifter: it combines elements of diverse media technologies and institutions. This has implications for the analytical frameworks we use in media research. The trick is not to take an either/or approach, trying to shoehorn Netflix into one box or another, but rather to see it as a media object that *performatively enacts* its association with these media at different times and for different purposes. In its dealings with government, Netflix claims to be a digital media service—certainly not television, which would attract unwelcome regulation. Yet, in its public relations, Netflix constantly refers to television, because of its familiarity to consumers. Its interface design, on the other hand, prefers to evoke the cinema experience. Meanwhile, its subscription business model has echoes of pay-TV, but its algorithmic recommendation system is pure new media. In other words, Netflix is a hybrid technology that remediates a range of earlier media technologies in different aspects of its operation, and this mix of associations is constantly changing.

The good news for television studies is that these issues are already quite familiar to scholars. Television is a hybrid medium that combines and rearranges elements of previous media forms, including radio, cinema, newspapers, and the theater. Equally, television studies—to the extent that it exists as a discrete academic field—has evolved as a historical amalgam of different critical approaches, research methods, and ways of knowing. Television studies is a malleable discipline, and this natural flexibility will be an asset as we enter further into an era of internet-distributed television services, which requires us to keep an open mind as to what exactly television *is* and how it might be studied. In this respect, Netflix is an important object lesson precisely because it invites us to revisit what we think we know about television and to reconstitute that knowledge anew.

Arguably, what is more important than what we call Netflix is how we think about it. In this chapter, I have argued for a both/and perspective, suggesting that we should acknowledge the specificities of Netflix as a digital media service (such as its mode of interactivity, algorithmic filtering, and regulatory slipperiness) and what this means for its distribution function (its catalog structure, lack of capacity limitations, and nonlinearity) while also appreciating the continuities between Netflix and broadcast media, which are especially noticeable at the level of text, engagement, and experience (the "it's still TV" argument). It is not enough to treat Netflix just like any other digital platform, because this misses its specificity as a hybrid TV-cinema-digital media distribution system with a unique set of experiential and aesthetic connections to older media. Nor is it enough to wheel out the standard theories of television studies and apply them to Netflix. A better approach would be one that is literate in both screen and digital media studies and can move between these ways of knowing. The need for such an approach will become evident in the next two chapters, when we turn our attention to Netflix's distribution model and infrastructure.

2

Transnational Television

From Broadcast to Broadband

There are few issues in contemporary television studies that cannot be traced back in some way to the 1974 book *Television: Technology and Cultural Form* by Raymond Williams. Of particular interest for scholars of internet-distributed television is the book's final chapter, "Alternative Technology, Alternative Uses," which offers a richly textured account of new distribution technologies and their sociopolitical implications. Writing in the early 1970s, Williams could not have foretold the rise of Netflix. Nonetheless, his discussion of emerging satellite television services identifies issues that are directly relevant to today's debates about transnational television in the internet age.

Williams viewed satellite television as a site of structural conflict—between competing institutions, business models, and visions of what television is and should be—as well as being a staging ground for Cold War politics. He was especially interested in the transnational dimension of satellite distribution and what this might mean for global communication. Noting on the one hand that

Figure 2.1. Video platforms, including YouTube, operate transnationally but are territorialized through geolocation and personalization. Photo by Kapustin Igor/Shutterstock.

satellite's "probable" evolutionary trajectory would be to "penetrate or circumvent existing national broadcasting systems, in the name of 'internationalism' but in reality in the service of one or two dominant cultures" (Williams 1974, 147), Williams offers a highly ambivalent assessment of the forms of television that may result from satellite distribution:

> A world-wide television service, with genuinely open skies, would be an enormous gain to the peoples of the world, as short-wave radio, bypassing national controls, has already clearly been. Against the rhetoric of open skies, which in fact, given the expense and sophistica-tion of satellite technology, would be monopolised by a few large corporations and authoritarian governments, it will sound strange and reactionary to defend national autonomy. But the probable users of the technology are not internationalists, in the sense of any significant mutuality. The national or local components in their services would be matters merely of consent and pub-licity: tokenism. (149)

In this quote, we can observe several clashing ideas that continue to structure today's debate about internet-distributed television. On the one hand, there is the utopian vision of a "world-wide television service," seen here as a potential global agora—a space of free and recip-rocal exchange. On the other hand, there is the recognition that this space is likely to be organized around existing forms of industrial and geopolitical power; hence the cos-mopolitan space of transnational communication also becomes a space of domination. Finally, there is Williams's qualified appeal to national regulation as a bulwark against multinational corporations. Using language reflecting 1970s debates about cultural imperialism, Williams warns that the advent of commercial direct-to-home satellite television systems may make independent production "very difficult or impossible" and that most "inhabitants of the 'global village' would be saying nothing ... while a few powerful corporations and governments, and the people they could hire, would speak in ways never before known to most of the peoples of the world" (149).

It is not difficult to see the links between Williams's Cold War–era predictions of cultural imperialism from above and current fears about U.S. cultural domination in internet-distributed television services. More than 40 years after the publication of Williams's book, we still do not have a single "world-wide television service"—a distribution system or platform that is widely accessible in every part of the world. We do, however, have a range of *transnational* multiplatform television services—including international news channels (CNN, Al Jazeera, Russia Today), internet-distributed subscription services (Netflix, Amazon Prime Video), and online video-sharing platforms. Each of these services has its own underlying technologies, distribution patterns, and ways of reaching dispersed markets. They are all transnational but *in different ways*, and just as Williams predicted, many of these services have become controversial because of the way they impact national markets, allegedly reshape national cultures, and evade national regulations.

This chapter asks: What is distinctive about the transnational character of internet-distributed television compared to earlier forms of transnational television? In answering this question, it seeks to locate current debates in a wider historical context. While we often think about digital media in a vacuum, as though each new innovation was the first of its kind, many of the concerns about Netflix and other transnational internet-distributed television services have clearly been raised before—including the fear of cultural penetration by powerful nations, the weakened power of the nation-state, the lack of local content, and the privatization of public institutions. With these issues in mind, the present chapter will describe key structural changes in television distribution since the 1970s and explain how today's multiterritory SVOD services appear when seen through the lens of historical debates.

From National to Transnational Television— and Back

"Transnational television," as I use the term here, refers to the propensity for television distribution systems to cross one or more national borders. It is a deceptively simple term that invokes a wide range of scenarios, including both cosmopolitan and culturally intrusive distribution. For our purposes, the related term "global television" will refer to television services that operate in a large number of international markets simultaneously. Netflix, by my definitions, is both transnational and global. HBO, in contrast, is transnational but not global, because it offers its standalone internet-distributed service (HBO Now/Go) only in select markets in Latin America, Central Europe, and Asia. Most national catch-up services are neither transnational nor global, at least from the point of view of distribution.

The history of broadcast television is closely tied to the history of the nation-state. Since the interwar period, the organization of television systems in almost every country has mirrored and indeed reinforced national boundaries. The nationwide distribution of television has shaped advertising markets, has propagated official language policies, and has established common frames of national discourse. As Jean K. Chalaby writes,

> For much of its history, television has been closely bound to a national territory. Broadcasters exchanged programmes and set up international associations, but operated within national boundaries. Their signal covered the length and breadth of the country, from the nation's capital to the remotest parts of the countryside. Foreign broadcasters were not allowed to transmit on national territory and attempts to do so were seen as breaches of sovereignty. Television was often tied up with the national project and no other media institution was more central to the modernist intent of engineering a national identity. (Chalaby 2005, 1)

These institutional contexts produced a particular industrial structure. Distribution was contiguous with territory, and control over television institutions rested clearly (though not always easily) with national governments. Regulation ensured a national "container" around television, creating markets, institutions, and viewing cultures that aligned predominantly with national borders.

Since the 1970s, successive technologies have undermined this structure, complicating television's spatial dynamics. Satellite and cable distribution have taken television signals into new places. Digitization and compression have increased the number of channels. Successive waves of liberalization have swept through the international TV system, leading to the privatization of state broadcasters, and the deregulation of infrastructure, advertising, and content controls. Contemporary television industries are now characterized by dense, overlapping palimpsests of technologies, markets, and viewing habits, none of which are easily contained within national borders.

As a result, scholars are increasingly conceptualizing television as both a national *and* a transnational technology. Since the 1990s, a rich body of descriptive and theoretical literature has emerged. Chalaby (2005, 2009) has analyzed the industrial logics of satellite channels. Lisa Parks (2005) has studied the infrastructural and cultural dimensions of satellite television. Naomi Sakr (2001) has examined satellite television cultures in the Middle East. Richard Collins (1993, 1998) has analyzed transnational television policy in Europe. Other scholars have looked at the transnational reception of texts, genres, and formats (Katz and Wedell 1977; Gillespie 1995; Iwabuchi 2004; Moran 2009; Chalaby 2016). Reading across these studies, the nomenclature of the transnational appears to represent an attempt to come to terms with a variety of interlocking issues: the cross-border *mobility* of television content, talent, and formats; the *interaction* of international broadcasters, regulators, and institutions; and the *cosmopolitanization* of television audiences, styles, and viewing habits.

A related body of scholarly literature on global television has also emerged. Sometimes this term is used interchangeably with "transnational television" or as an umbrella term for "all the world's television" (Barker 1997). It may also refer to a particular *scale* of operation—commonly associated with channels such as CNN or BBC World, or platforms such as YouTube and Netflix, that are available in a large number of countries simultaneously. Alternatively, "global" may evoke a particular critical epistemology. For example, Lisa Parks and Shanti Kumar see the rubric of global television as an opportunity to rethink the objectives of television studies in a cross-cultural frame:

> How do we write in a way that captures the movement of television programs across national borders and cultures? How do we describe the unequal technological access and television production around the world—what could be called global television's *ebbs and flows*? How can we describe television in places where we do not speak local languages? How do we study transnational audiences, viewers scattered across continents? How did certain television institutions and industries emerge in different parts of the world? (Parks and Kumar 2003, 3, emphasis in original)

This passage was written before internet distribution was a mainstream feature of television, but its relevance to contemporary concerns is clear. The themes of "movement," "ebbs and flows," translation, access, "scattered" audiences, and patterns of emergence, central to the television studies agenda outlined by Parks and Kumar, are also central to the account of digital distribution offered in this book.

Clearly, there are many aspects of internet-distributed television that are unambiguously transnational, if not global: the simultaneous release of content across dozens of national markets, the ability to instantly stream material from faraway countries, the rise of online translation networks, and so on. But we must also pay attention to the ways in which internet-distributed television is bounded, "placed," and restricted, and how its patterns of inclusion and exclusion are spatially organized. Rather than conceptualizing the relationship between these scales in a dialectical way (the national versus the transnational or global), the best way forward may be to develop more finely honed analyses of the different kinds of mobility enabled by television distribution technologies and their associated market forms. After all, what we call television is in fact a bundle of different technologies and practices, each with its own spatial characteristics, its own ways of crossing borders. To understand what transnational television means in the internet age, we need a more granular account of the key distribution technologies in this bundle, including broadcast, satellite, and internet distribution. We begin with broadcast.

Spatial Logics of Television Distribution

From a technical perspective, all broadcast media are potentially transnational in the sense that radio waves are multidirectional. Radio transmission may be limited by topography (mountains and skyscrapers block out signals), but it is not limited, in a technical sense, by political boundaries. Television as a broadcast medium has always had a transnational dimension because radio waves do not respect borders. While most broadcasting takes place within, rather than across, national borders, there is nothing inherent in the technology that says this must be the case; indeed, the history of broadcasting is full of examples to the contrary.

Radio—television's predecessor—was an international medium well suited to crossing borders. Michele Hilmes (2012, 2) claims that "no previous medium possessed the equally important capacity of radio waves to *transgress* national borders, to defy barriers of both time and space, to travel unseen through the air and enter the ears of private citizens in their homes, undetected by public gatekeepers" (emphasis in original). Since the earliest experiments with cross-border broadcasting at the turn of the twentieth century, entrepreneurs, enthusiasts, and government officials have all found ways of using radio to reach faraway minds and markets. This inherent transnationalism was the reason that governments all over the world sought to strongly regulate radio or to use it as their own propaganda channel (Hilmes 2012).

With its ultralong transmission range, shortwave radio was the favored medium for government propaganda broadcasting during the Cold War. Radio Free Europe penetrated deep into the Eastern Bloc, while Radio Moscow could be heard in Africa, the Middle East, Asia, North America, and Western Europe until the 1980s. Creatures of national government, these were *transnational* media from the point of view of distribution. There are also many examples of commercial radio stations operating transnationally, usually without authorization. Radio Luxembourg (later Radio Television Luxembourg, or RTL) is the most famous case. In the 1930s, it used its transmitter—then the most powerful in the world—to beam out multilingual services across much of Western Europe. In the 1960s, pirate radio stations such as Radio Caroline targeted British listeners from international waters, breaking the monopoly of the BBC and breeding new youth cultures (Johns 2011). The U.S.-Mexico frontier was another popular site for transnational broadcasting: "border-blaster" AM radio stations in Mexico reached large audiences in the southwestern United States, interfering with the signals of local channels and attracting large youth audiences, as noted by Burroughs (2015), who has analyzed these continuities between transnational radio and digital streaming.

A common distribution logic of transnational broadcasting is *signal spillover*. Wherever two nations share a border, there is often accidental reception of TV signals on one or both sides. European markets, with their dense patchwork of languages, cultures, and media systems all jostling for territory, are particularly susceptible to spillover. Many Europeans can choose between broadcast services based in their own and neighboring countries—Belgians, for example, can access Dutch, French, and German broadcast television channels. The politics of spillover were heightened during the Cold War, when East Germans watched West German channels, Hungarians watched Austrian channels, and Albanians watched Italian and Yugoslav channels (Jakubowicz 1994). Similarly, in Canada, spillover from U.S.-based media is an everyday fact of life: most Canadians live near the U.S. border, thereby constituting an unofficial advertising market for U.S. television stations. Africa also has its own patterns of spillover, related to uneven geographies of development. In the northern part of Tanzania—which did not have a national television service until the 1990s—people have long been able to watch Kenyan television, while audiences in Botswana and Lesotho similarly enjoyed South African television broadcasts long before they had their own national TV channel (Mytton, Teer-Tomaselli, and Tudesq 2005, 97).

These examples provide a modest correction to the accounts of broadcast television that seek to characterize it as essentially a national medium. While the institutional history of broadcast television is a history of nationally

defined regulation, markets, and audiences, its technological and social history (how signals traveled and how
audiences were formed) is full of unofficial and unavoid-
able border crossings. These leakages generally occurred
through spillover or deliberate targeting of foreign audiences, as in the case of RTL. Given the limited reach of
broadcast distribution, it is more appropriate to call these
transnational rather than *global* media. Aside from shortwave radio, with its ultralong-range capacity, most forms of
transnational broadcast media involved localized leakages
across one or at most a few national borders.

With the advent of satellite technology, the transnational aspect of television increased significantly. A broad-
beam satellite's footprint can cover up to 40% of the earth's
surface, providing signals to an unlimited number of re-
ceivers and overcoming geographical obstacles, including
mountain ranges and oceans. Satellite therefore entails a
different reception model from broadcast television, as
well as a different political economy, characterized by high
start-up costs but formidable geographic reach. As Lisa
Parks writes, "satellite television practices . . . have helped
to determine (that is, to shape and set the limits of) the
spheres of cultural and economic activity that constitute
what we know as 'the global'" (Parks 2005, 2).

The history of satellite TV can be divided into two overlapping periods. Beginning in the 1960s, communications
satellites were used as a business-to-business distribution
technology enabling retransmission of signals (Straubhaar
2007). The Russians were the first to develop a nationwide
satellite TV system, using the Molniya satellites in 1967.
In the United States, HBO and Ted Turner's WTBS used
satellite feeds to relay programming to their partners
(thus removing the need to courier videotapes between

stations). These applications changed the industrial logic of television by accelerating the development of particular formats, genres, and modes of address. Direct transmission of programming from big-city TV studios to regional affiliates led to the rise of national networks, and the ability to beam content to international partners created global made-for-TV spectacles such as Our World and Live Aid. In the 1980s, with the advent of satellite TVRO (TV receive-only) and, later, direct broadcast satellite TV, satellite evolved into a domestic consumer technology. Millions of people installed satellite dishes on their rooftops and decoder boxes in their living rooms, many of them tuning in to transnational channels such as CNN, Sky TV, Music Box, and MTV.

Satellite was (and still is) a controversial technology. A driver of competition, a form of cultural trespassing, an affront to national sovereignty, a vehicle of marketization, an accidental nation-builder—satellite was all these things and more. The key political debates of the satellite age were about whether signals from "foreign" countries should be allowed "into" national space. Views on this topic varied greatly, of course, depending on the countries in question, their fear of cultural domination, and the industry stakeholders involved. Spatial metaphors—trespassing, intrusion, spillover, and leakage—ruled the debate.

The impact of satellite technology on national communications policies was profound. Bitter debates about satellite technology's effect on national media sovereignty raged across the world, including at the United Nations, UNESCO, and the International Telecommunication Union. On one side, advocates of the new communications order championed the emancipatory possibilities of transnational communications. Liberal social scientist and futurist Ithiel de Sola Pool was particularly enthusiastic, noting satellite technology's power to engender "multilateral flows" (de Sola Pool 1990, 70) and celebrating "that extraordinary march of progress which modern technology has brought" (129). For de Sola Pool, the key to satellite's magic was its border-crossing qualities:

> With a satellite, the communications distance between all points within its beam has become essentially equal. In the non-Euclidean communication plane that results, no point lies between two other points. Boundaries partition nothing. The topology of commerce, government, and social life may change to reflect that space warp. (65)

Yet there were many detractors. Leftist scholars warned of the potential for "trespassing over national boundaries on an unprecedented scale" (Nordenstreng and Schiller 1979, ix). Ploman (1979, 155) predicted "an unregulated flood of satellite-born television programs . . . that could impair cultural identity and media policy in the 'receiving' countries." Raymond Williams lamented the prospect of satellite TV becoming a pretext for "penetration by para-national corporation advertising" (Williams 1974, 149).

A classical anxiety provoked by satellite television concerns unwelcome market entry, or the capacity for satellite-delivered channels to unfairly compete with national media companies. Satellite distribution redraws market boundaries to include new players so that incumbents with decades of history in a national market are suddenly in competition with powerful foreign operators—companies that have not "paid their dues" to the government through licensing fees, regulatory adherence, and political patronage. Forty years later, this unfair competition argument is staging a comeback in contemporary debates about SVOD, as local operators bemoan the unwelcome presence of Netflix and Amazon Video in their home markets.

In the 1990s, a series of books by influential media scholars pointed to other, more subtle effects of satellite technology on national television markets. In *Spaces of Identity* (1995), David Morley and Kevin Robins wrote eloquently about satellite television policy in Europe during the 1980s and 1990s—a period of intense debate about the European Union's plans for liberalization and unification of its member states' audiovisual industries, laws, and regulations. Its cornerstone policy, *Television without Frontiers*, adopted in 1989, emphasized the free movement of European content across the region. In their book, Morley and Robins provided a rich theoretical account of the changing spatiality of television in Europe and the new kinds of media geography thus created, emphasizing the "restructuring of information and image spaces and the production of a new communications geography, characterised by global networks and an international space of information flows" (Morley and Robins 1995, 1):

> There is, then, an expansionist tendency at work, pushing ceaselessly towards the construction of enlarged audiovisual spaces and markets. The imperative is to break down the old boundaries and frontiers of national communities, which now present themselves as arbitrary and irrational obstacles to this reorganisation of business strategies. Audiovisual geographies are thus becoming detached from the symbolic spaces of national culture, and realigned on the basis of the more 'universal' principles of international consumer culture. The free and unimpeded circulation of programmes—television without frontiers—is the great

ideal in the new order. It is an ideal whose logic is driving ultimately towards the creation of global programming and global markets—and already we are seeing the rise to power of global corporations intent on turning ideal into reality. (11)

Discussing some apocryphal moments of 1990s-era globalization rhetoric,[1] Morley and Robins emphasized the difficult interactions between transnational media and the tastes, values, and practices of actually existing audiences. "The question that we must now consider," they wrote, "is how this logic unfolds as it encounters and negotiates the real world, the world of already existing and established markets and cultures" (15). What Morley and Robins were pointing to here was the scalar tension between the global and the local, and what happens when the former meets the latter. This remains a fundamental tension within television distribution, as the current controversies about Netflix and local content suggest (see Chapters 4 and 5).

Around the same time that *Spaces of Identity* appeared, John Sinclair, Liz Jacka, and Stuart Cunningham published *New Patterns in Global Television: Peripheral Vision* (1995). Sinclair, Jacka, and Cunningham placed special emphasis on satellite's power to create regional geolinguistic markets (e.g., a transcontinental Spanish-language market) rather than simply erasing national differences and instituting a flat plane of global flow. They noted the structural impacts of satellite technology for the wider political economy of television:

There is no doubt that the satellite has acted as a kind of "Trojan horse" of media liberalization. Although evidence from Europe and elsewhere indicates that satellite services

originating outside national borders do not usually attract
levels of audience that would really threaten traditional
national viewing patterns, the ability of satellite delivery
to transgress borders has been enough to encourage generally otherwise reluctant government to allow greater
internal commercialization and competition. (Sinclair,
Jacka, and Cunningham 1995, 2)

The authors identified a further consequence of satellite television, which was to vastly increase channel
capacity. This had the effect of stimulating demand not only
for imported Hollywood content but also for "programmes
from new sources, including some formerly peripheral
regions," including Australia, Canada, Latin America, India,
and the Middle East (3). They argued that satellite distribution, as well as expanding and liberalizing media systems,
was also creating capacity for what Thussu (2006) later
called "contra-flow."

Joseph Straubhaar, in his wide-ranging study *World
Television: From Global to Local* (2007),[2] identified further
contradictory effects of satellite technology. The first was
to "permit complete national distribution and penetration
of television" (Straubhaar 2007, 124) in large countries such
as China, Canada, Brazil, and the former USSR, thus paradoxically reaffirming the centrality of the nation by bringing its territory into the same satellite footprint. A second
effect was to foster the growth of global news and entertainment channels and niche services targeting diasporic
audiences. Straubhaar also noted a third effect, reflecting
"a quite different logic of globalization" (125), which was
to hook elite classes directly into regional and global flows
(for example, by letting middle-class viewers in Latin

America watch global channels like MTV, thus signaling their difference from the masses who prefer local channels). In an important qualification to previous research, Straubhaar argued that satellite's spatial effects were not just about facilitating transnational flows. Satellite distribution also had nation-building effects.

What can we learn from these revisionist accounts of satellite television? The key point is that disruptive distribution technologies such as satellite often turn out to have diverse and unexpected effects on national television markets. A frequently observed feature of satellite distribution was that it enabled long-distance market entry for transnational broadcasters. However—as Straubhaar and Sinclair, Jacka, and Cunningham noted—there were also countervailing tendencies to build national and regional markets. Satellite, while certainly a transnational technology in terms of its inherent technological affordances, also worked to *integrate* national and regional spaces.

We can see similar things happening with internet television today. For example, there is a strongly national dimension to many portals, especially catch-up and AVOD services offered by public-service broadcasters, commercial networks, and pay-TV companies, which tend to provide on-demand redistribution for content that is already available or familiar to local audiences, sometimes with a few digital exclusives thrown in. This is the model used by many portals, whether CBS All Access in the United States, ABC iView in Australia, or Blim in Mexico. Inherently transnational from the point of view of its distribution technology (IP distribution), these platforms acquire a strongly national character in their institutional form. This kind of internet-distributed television is not generally in the business of introducing new transnational flows that were not already a feature of their home markets. Its aim is to extend an existing paradigm of television viewing into the digital space and provide *more of the same* to a nationally defined audience.

Other internet television services have more explicitly transnational and transcultural effects. SVOD services for cinephiles and genre fans (such as Mubi, Viki Pass, and Doc Alliance Films) tend to operate transnationally by aggregating small audiences in many nations, license terms permitting. YouTube is also highly transnational because of its open nature, catering to a very wide range of viewing communities.

Netflix, as a transnational service operating in many different markets simultaneously, is an unusual case. On the one hand, Netflix's over-the-top model resembles satellite and, to a lesser extent, broadcast distribution, in that it can penetrate national media spaces without state sanction. Netflix's arrival in international markets has therefore resurrected the specter of cultural trespassing—a spatial logic embodied in the term "over-the-top." For example, in January 2016, after the Netflix global switch-on, officials at the Kenya Film Classification Board issued a press release objecting to the unauthorized arrival of this culturally foreign service within the Kenyan media space. "The board regards this development as a gross contravention of the laws governing film and broadcast content distribution in Kenya," read the board's statement. "As a progressive country, we cannot afford to be [a] passive recipient of foreign content that could corrupt the moral values of our children and compromise our national security. . . . The pornography, child prostitution and massive violence themes in some of the movies threaten our moral values" (Rajab 2016; Aglionby and Garrahan 2016; Kuo 2016). Statements like this recall the satellite-era paradigm of cultural intrusion from above, in which a foreign media service encroaches into the sovereign space of the nation—unwelcome and uninvited.

Authorities in Russia have also objected to Netflix's unauthorized arrival using similar language. "Before entering the market," said Deputy Communications Minister Alexei Volin, "Netflix should have had consultations with Russian representatives, including regulating agencies" (Volin in Kozlov 2016). Russian culture minister Vladimir Medinsky (cited in Borenstein 2016) went somewhat further in a June 2016 interview, stating that, "Our ideological friends [the U.S. government] are well aware what constitutes the most important of all art forms [cinema, according to Vladimir Lenin], and they understand how to enter everyone's homes by getting into every television with the help of Netflix. And through this television, [they get into] the heads of everyone on Earth." In 2017, Russian regulators passed a law limiting the foreign ownership of major streaming services to 20%—thus effectively banning many U.S. services from operating on Russian soil, or obliging them to team up with Russian media companies in joint ventures. This regulatory model is an extension of laws governing Russian media companies, which had previously led to the exit or forced restructuring of U.S.-owned corporations doing business in Russia. At the time of writing, Netflix is still operating in Russia, because of a provision in the law that gives a free pass to smaller companies with under 100,000 daily viewers. Netflix—which has not commented publicly on the law and never reveals national subscriber numbers—is thought to be under the threshold (Kozlov 2017).

How well founded are these regulators' concerns, given Netflix's distribution model? Just because Netflix was suddenly switched on in Russia and Kenya, does this mean that it is now present in the same way as a broadcast, satellite, or cable channel? What degree of cultural power can be ascribed to an internet-distributed service as it enters a new market? How governments and citizens feel about "foreign" media is fundamentally dependent on the origin, target, and cultural context of the content being transmitted. Shortwave radio broadcasts were welcome from allies but unwelcome when they came from enemies. Satellite beams can diversify national media landscapes or arrogantly trespass on sovereign territory. Similarly, transnationally distributed internet television can be seen by governments as *beneficial* when reconnecting communities with national discourses, *benign* when circulating "mere entertainment," and *threatening* when undermining national media regulation or introducing unwelcome ideas into the national media space.

Given that the political discourse around Silicon Valley tech companies is changing rapidly at the moment, over the next few years it will be instructive to follow how Netflix and other internet television services are construed as allies or enemies in various national contexts. This is a highly volatile area for media policy. Just as the history of satellite television cannot be divorced from the wider horizon of Cold War politics, cultural imperialism debates, and economic liberalization, contemporary debates about internet television must be seen in light of monopoly fears, local-content protectionism, tax-evasion scandals, and the general policy backlash against big tech.

Rethinking the Transnational

It is difficult to compare the degree of transnationalism inherent in different technologies of television distribution, even though we can see clear differences in affordances. Internet-distributed television is both *more* and *less* transnational than older television systems, depending on the criteria you use to measure such things. One of the challenges here is that we lack an adequate vocabulary to describe the geographical configurations characteristic of internet-distributed television. We may need to go beyond terms like global, national, and transnational; however, as yet there is no consensus regarding viable alternatives.

In his 1994 book *Virtual Geography*, Mackenzie Wark used the *vector* as the appropriate spatial metaphor for describing the geography of emergent media. Borrowing from Paul Virilio, Wark describes the vector's ability to "link almost any points together" and to "connect enormously vast and vaguely defined spaces together and move images, and sounds, words, and furies, between them" (Wark 1994, 11–12). Although Wark was writing about transnational television events of the 1990s (including satellite-distributed CNN broadcasts of the first Gulf War) rather than the internet, his formulation influenced a generation of cyberculture theorists, resonating with the idea of global connectivity found in manifestos such as Nicholas Negroponte's *Being Digital* (1995). Other metaphors have been offered by Rob Kitchin and Martin Dodge in their book *Code/Space* (2011) and, most famously, by sociologist Manuel Castells in his account of the "space of flows" in *The Rise of the Network Society* (1996) and the "Internet galaxy" in *The Internet Galaxy* (2001).

Other scholars have questioned the possibility of ascribing a single spatial metaphor to internet distribution. As David Golumbia notes, even terms such as "centralized" and "decentralized" have conflicting meanings when applied to internet phenomena:

> Facebook, for example, might be seen as decentralized because it is made up of its millions of users, spread out all over the planet; as centralized, because one company collects all of the data from those users; as decentralized, because all that data is not housed in a single geographic location but on servers all over the world; as centralized, because those locations are nevertheless tightly held together via software and hardware; and so on. (Golumbia 2016, 51)

As Golumbia shows, the weak applicability of analog-era spatial concepts to the internet presents a problem for critical analysis. Matthew Zook (2006, 54) similarly observed that the internet has no singular geography, only multiple economic, political, and human "geographies . . . created through the interaction of this technology with the places in which and the people by whom it is used."

Clearly, it is a challenging task to describe the spatial logics of internet distribution in the same way that we have described broadcast and satellite distribution. Notions of a signal zone or footprint do not translate well into this paradigm. To make things more complicated, each internet-distributed service has its own unique spatial pattern because of the interactions of variables such as geographic availability, content licensing terms, user base, pricing, and so on. There are also important differences in business models that shape the geographic availability of individual platforms (see Table 2.1).

TABLE 2.1. Examples of national, multiterritory, and global internet-distributed TV services as of January 2018

	National single-territory	**Transnational** multiterritory	**Transnational** global
AVOD[a]	SBS On Demand (AU)	Crackle (21 countries)	YouTube[d] DailyMotion
TVOD[b]	UniversCiné (FR) Maxdome (DE)	RakutenTV (11 countries) Microsoft Movies & TV (22 countries)	iTunes Google Play[e]
SVOD[c]	Blim (MX) CraveTV (CA)	HBO Now/Go (43 countries)	Netflix Amazon Prime Video

[a] AVOD: Advertising-based video on demand.
[b] TVOD: Transactional video on demand.
[c] SVOD: Subscription video on demand
[d] YouTube, while predominantly used as a free AVOD service, is technically a hybrid service, which also includes SVOD (YouTube Premium) and transactional purchases.
[e] Movies are available through Google Play globally, though TV shows are presently only available to customers in Australia, Austria, Canada, France, Germany, Japan, Switzerland, the United Kingdom, and the United States.

Netflix is an interesting case because the company has so proudly and loudly proclaimed its global status. "You are witnessing the birth of a global TV network," said Reed Hastings at CES in January 2016. "With the internet, global distribution no longer needs to be fragmented. It means that everyone pretty much everywhere should be able to see great films and TV shows at the exact same moment," he added. Global simultaneity is certainly a striking feature of Netflix, when seen in the context of the longer history of sequential film and television distribution (windowing). There are also other attributes that make Netflix a distinctly transnational, even cosmopolitan, service. It hosts a very diverse, though Hollywood-centric, spread of multilingual content, and it now translates its original content into dozens of languages. By aggregating large amounts of content into the one platform, Netflix also enables a certain cosmopolitan consumption experience that is highly valued by some subscribers. Some of my favorite Netflix memories involve stumbling on television series and telemovies from the far reaches of the catalog that would otherwise be inaccessible in Australia—from Russian gangster shows to Turkish soap operas. In this sense, Netflix can rightly be considered transnational, global, and cosmopolitan, at least to some degree.

But Netflix is also fundamentally *national* in several important ways. Headquartered in the United States, it has largely projected and expanded its existing business model and philosophy of entertainment into new countries (while localizing itself through licensing and original production). In this sense, one could argue that it is more multinational than transnational. Conversely, its reliance on territorial copyright licensing means that it may best be understood as a series of national media services stitched together into a single platform. It seems that Netflix does not fit particularly well with the scalar vocabulary of national/transnational/global on which the field of media studies has traditionally relied. As a digital service, it takes elements from each of these scales and combines them in new ways.

Perhaps the best way to describe Netflix would be to follow Golumbia's lead and simply note the contradictions. Netflix was born global in the sense that it is an internet-distributed service, but it is highly territorial because of its licensing model. It is quasiglobal in its *reach* (available everywhere but China, Iran, North Korea, and Crimea), but its actual uptake and use is thus far concentrated in a few regional markets (Western Europe, the Nordic countries, Latin America, and the Anglosphere). It is centralized in the sense that its headquarters and its ways of understanding the world are firmly located in California, but it is also decentralized at a technical level through its content delivery architecture (see Chapter 3). In other words, Netflix simultaneously *reflects*, *embraces*, and *resists* the possibilities for transnational distribution that are inherent in internet-distributed services.

To put this differently, Netflix is global but is not a "wraparound"; it does not evenly envelop the world. It has wide reach but narrow impact. Perhaps it may be better to imagine Netflix as a kind of loose mesh, one that is full of gaps, that comes into contact unevenly with local and national surfaces; or, alternatively, as a cluster of network trajectories—similar to shipping route maps—that cohere around preexisting concentrations of connectivity and capital. These distinctions are to some degree semantic, but they point to a larger problem for media theory, which is that current vocabularies for describing space and scale in digital media may no longer be fit for purpose. Part of the job of global television research over the next decade should be to invent alternative analytical vocabularies that might better reflect this changing landscape.

3

The Infrastructures of Streaming

Internet television promises instant access to content at the touch of a button or the swipe of a finger. Behind this apparent simplicity lies enormous technical complexity. Once we begin to look into the back end of a service like Netflix, we get a sense of the interlocking systems that are essential to the experience of digital media but that are hidden from view in everyday usage.

Consider for a moment everything required to deliver a Netflix stream to a user located in Singapore or Santiago. An inventory of these systems (which would be many pages long) might include the viewing devices, modems, routers, and other consumer hardware that enable users to connect to internet television services (and their associated programming languages, protocols, and technical standards); the telephone lines and fiber-optic cables that carry voice and data traffic to the home; the content delivery networks (CDNs) that cache video content in servers near end users; customer management software and third-party payment processing systems; and of course the power grids and undersea cables that make all this activity possible. Internet television never *just* works but must be *made* to work, through a vast complex of

Figure 3.1. The Netflix "hourglass" icon that appears while a stream is loading. The length of the delay depends partly on the user's internet connection speed. Screenshot by the author.

engineering, maintenance, pipes, pits, and governance—in short, infrastructure.

Infrastructure, Paul Edwards writes, is "the invisible background, the substrate or support, the technocultural/ natural environment, of modernity" (Edwards 2003, 191). Ostensibly hidden from view, at least until something goes wrong, infrastructure is a challenging topic for contemporary media studies. This chapter, while not a comprehensive technical account, provides some starting points for understanding Netflix from an infrastructural perspective. In so doing, it poses a number of conceptual questions: What happens to our ideas about internet television when we look not at the content or interface but at the underlying systems that deliver video to users? How are these systems spatially organized, and how do content and data move across them? What might all this add to our understanding of global media?[1]

The Infrastructural Optic

As the first step, let us revisit some conceptual debates surrounding infrastructure as they have been taken up in media studies. In recent years, scholars have become increasingly interested in the invisible networks, systems, and standards that underlie our everyday media experiences. Media infrastructure is now becoming a rich area of study in its own right, as evidenced by the publication of books such as Jonathan Sterne's cultural history of audio compression, *MP3: The Meaning of a Format* (2012), and the anthology *Signal Traffic: Critical Studies of Media Infrastructures* (Parks and Starosielski 2015), a collection of essays on data centers, mobile phone towers, and e-waste. Popular trade books, such as Andrew Blum's

Tubes: A Journey to the Center of the Internet (2012), are also opening people's eyes to the scale and complexity of international telecommunications infrastructure.

Parks and Starosielski, writing from a screen and visual studies perspective, define media infrastructures as "situated sociotechnical systems that are designed and configured to support the distribution of audiovisual signal traffic" (Parks and Starosielski 2015, 4). Other scholars understand infrastructure in a more expansive sense. Urbanist Vyjayanthi Rao, for example, describes infrastructure as "the organizational medium of urban life" (Rao 2014, 39). Notwithstanding definitional differences, what is exciting about this turn to infrastructure in critical humanities and social science is that it invites engagement with topics that were previously out of bounds, or at least inaccessible, for many humanists—issues related to electrical engineering or information systems design, for example. It also reflects a new intellectual curiosity: a desire to use infrastructure as "an analytic and a research method" (Sandvig 2015, 91). The infrastructural turn involves not only new objects of analysis—fiber-optic cables, data centers, compression technologies, standards, internet routing protocols—but also new ways of seeing and narrating those objects and connecting the resulting discussions to critical and theoretical debates.

The roots of the infrastructural turn in media studies can be traced along different lines. Pioneering scholars of communication, including Harold Innis (1951) and Ithiel de Sola Pool (1990)—both now rediscovered by a new generation of media researchers, though in different ways—have long drawn our attention to the materiality of communications technologies and their far-reaching connections to land, state, and empire. Looking further afield, there is the enduring influence of science and technology studies (STS)—especially the work of Bruno Latour, John Law, and their colleagues—which is concerned with the interaction between users, systems, standards, metrics, and other nonhuman actors. The "infrastructure studies" movement in the United States, a loose grouping of historians, social scientists, and information researchers, has produced many engaging works on technical systems and standards (Edwards 2003; Edwards et al. 2007; Star and Ruhleder 1996; Star and Bowker 2000; Lampland and Star 2008). There is also a tradition of geography of telecommunications that emphasizes the situated and material aspects of communication networks (Graham and Marvin 1996; Warf 2013).

While all of these approaches are different, together they provide powerful concepts for thinking about infrastructure and how it shapes communication. The key ideas can be summarized as follows:[2]

- *Reliance* Digital media would not exist without both "hard" infrastructure (for electricity, lighting, and telephony) and "soft" infrastructure (programming languages, standards, and protocols). These are the preconditions for the digital media experiences we take for granted (Abbate 1999; Blum 2012).
- *Invisibility and breakdown* Infrastructures "reside in a naturalized background, as ordinary and unremarkable to us as trees, daylight, and dirt" (Edwards 2003, 185). In other words, we only tend to notice infrastructure when it breaks or slows down. In digital media, the materiality of infrastructure often surfaces through the user experience of dropouts, slow loads, freezing, pixilation, and missing subtitles (Larkin 2008; Jackson 2013).

- *Codetermination* Infrastructure shapes communication and vice versa. Infrastructures are more than dumb pipes through which content travels. They play a role in shaping the experience of media, in the sense that the capacities and limitations of infrastructure are built into the system (as when online video services autoadjust their resolution to account for users' broadband speeds) (de Sola Pool 1990; Braun 2015).

- *Layering* Infrastructure builds on other infrastructure, resulting in palimpsests of interdependent systems. The geography of the internet, for example, adheres to the preexisting geography of telephone lines, which in turn follows the telegraph network. This results in the "layering or bundling of distinct systems" (Parks and Starosielski 2015, 9; Jackson et al. 2007).

- *Standardization* For infrastructures to work effectively, consensus is required as to the equipment, materials, processes, and formats used within it. This consensus is often the result of industrial conflict (as in the case of differing railway gauges, videocassette format wars, or battles over digital video and audio formats). Hence consensus is an outcome of competing interests and compromises, and must be actively maintained rather than taken for granted (Lampland and Star 2008; Sterne 2012).

These ideas provide a starting point for thinking about the infrastructure of internet television. When approached from this angle, Netflix—the world's biggest and most complex SVOD service, which distributes a billion hours of video content per week (Solsman 2017)—can be seen as the end product of interlocking and co-reliant technical systems, including a mix of public and private, open and closed, and soft and hard systems. Looking at Netflix from the perspective of these various systems (rather than from the platform or the content it hosts) produces a rather different image of how internet television works. For example, it might render visible the layering effects that result from decisions about how to regulate public communications infrastructure and whether the state should invest in data networks. It might draw our attention to the political economy of video standards and compression technologies, and how a platform's decision to use, say, Flash or Microsoft Silverlight is shaped by complex network effects and the dynamics of technological lock-in. Netflix, which previously used Silverlight for browser playback but switched to HTML5 from 2013 onward, was among the corporations that successfully campaigned for the inclusion of native digital rights management (DRM) capabilities within the open HTML5 standard—a move fiercely resisted by internet user groups such as the Electronic Frontiers Foundation. This shift away from a proprietary Microsoft plug-in made Netflix more accessible for some communities, including Linux users, but the flipside was that DRM was now built into the browser, a position Netflix strongly supported.

Another implication of approaching Netflix in this way is to bring into focus the *number* and *diversity* of infrastructures on which the service relies. From a software engineering perspective, Netflix—rather than being a single, monolithic architecture—actually relies on more than 700 microservices that run independently and talk to each other through APIs (application programming interfaces). Each microservice is programmed to do one specific thing—such as loading artwork for recommended titles or deducting the monthly fee from a customer's account (Nair 2017). Hence an infrastructural view reveals that Netflix is not really a singular platform; it is an ecology of small, purpose-built systems that work together to produce the effect of a singular platform.

A second implication here is that we need to pay attention to hard, soft, and human infrastructures simultaneously. Netflix relies on telecommunications and electricity infrastructures that provide energy and data routing through to the customer. It also relies on software-based infrastructures that are computational in nature but have their own material substrate. In addition, there is the human infrastructure of Netflix's programmers, customer service staff, marketing teams, and so on, not to mention its massive engineering workforce. As we move between hard, soft, and human infrastructures, socioeconomic variables become quite important in determining inclusion/exclusion dynamics. Hence it is not just a matter of Netflix requiring stable power to serve its customers (something that cannot necessarily be assumed in some developing countries). "Soft" infrastructures also require that we consider more subtle questions, such as: How does Netflix's reliance on credit card payment systems in most countries, rather than alternatives like cash and debit cards, work to exclude certain kinds of customers? How might the kind of operating system you use on your mobile device or set-top box shape your ability to use Netflix effectively? In other words, issues of infrastructural connectivity (power supply, bandwidth speed, etc.) are not the only things to consider in understanding Netflix's global reach. Payments, pricing, language availability, and various other matters are also infrastructural in nature and can be theorized infrastructurally alongside the pipes and cables.

Joshua Braun's (2013, 2015) work on digital video distribution explores these issues in depth. Braun, who works across the boundary of media studies and STS, argues that analysis of digital video services should not simply be an extension of existing paradigms of screen research but must also come to terms with the wide range of "transparent intermediaries" (Braun 2015) characteristic of internet technologies. "Unlike physical media, and their attendant icons of the paperboy or delivery truck," he writes, "we often have little intuitive sense of the route that electronic media take to get to us" (Braun 2013, 433). In an essay about Hulu and Boxee, Braun documents some of the more obscure intermediaries that now form part of the television distribution ecosystem:

> As video content wends its way to us online, it now goes through intermediaries most viewers have never heard of. Transpera (recently acquired by Tremor Media) for example, is a company that converts streaming video from numerous providers, ranging from Disney to CBS News, into a plethora of special formats tailored to our ever-growing menagerie of mobile devices, and packages advertising with it on its way to the consumer. YuMe is another company with major industry clients. It scans the blogs, homepages, and other sites on which users place embeddable videos and determines whether a page is "brand-safe" (i.e., that it features no objectionable content) before displaying paid ads with a clip. (Braun 2013, 433)

Braun's work shows how attention to the interlocking software systems that underlie digital media services can reveal a rather circuitous route between producer and consumer, changing our understanding of what distribution entails. Following John Law, Braun argues that an infrastructural approach to media research involves "a sort of archaeological interest in the various kinks, epicycles, and roundabouts found in a distribution route" that can ultimately "expose sociotechnical systems at work and lay bare the influence of infrastructure" (Braun 2015, 9). Following Braun's lead, we now turn to Netflix to see what this approach can tell us about the infrastructural geography of an SVOD platform.

Digital Divides and Download Speeds

Netflix is an internet-distributed service. As such, it relies on at least two different kinds of infrastructure: public and private telecommunications networks and its own internal networks and systems, including a bespoke CDN, which has a fascinating geography of its own. These two levels of infrastructure each have different affordances and spatial dynamics, and they interact to determine the kind of Netflix experience (or lack thereof) that users located in different parts of the world are likely to have.

Let us start with the geography of internet infrastructure: the mesh of fiber and coaxial cable, copper telephone wires, and satellite data links that form the internet's underlying foundation. While today's conversation about the digital economy often presumes a backdrop of constant connectivity, it remains the case that internet access is still unevenly distributed. As geographer Barney Warf writes,

> While those with regular and reliable access to the internet drown in a surplus of information—much of it superfluous, irrelevant, or unnecessary—those with limited

> access have difficulty comprehending the opportunities it
> offers, the savings in time and money it allows, and the
> sheer convenience, entertainment value, and ability to ac-
> quire data from bus schedules to recipes to global news.
> (Warf 2013, 2)

The location and reach of telephone cables; connection speeds and reliability; the variety of services available in particular locations; the relative takeup of home versus mobile internet and the use of cybercafés and public wifi; and the relations between communications infrastructure and urbanization are essentially spatial issues. Warf explains how access to the information revolution is not only unequally distributed but also spatially organized to include some and exclude others.

Sociospatial questions of access are the focus of what has become known as the digital divide debate. Initially focusing on the question of who is online and who is not, today's digital divide debate is also concerned with how different barriers to access interact and are compounded by variables such as class, age, gender, location, and education. It has shifted toward redefining access in "social as well as technological terms" (DiMaggio and Hargittai 2001, 3). As a result, today's conversations about digital inclusion and exclusion are increasingly about issues like the speed and reliability of connections, ISP pricing, mobile data allowances, interface design, consumer protection laws, and public wifi policies, as well as the geographic reach of telecommunications networks.

Consider how bandwidth limitations shape access to Netflix. The minimum bandwidth recommended by Netflix for a stable user experience is 0.5 Mbps (megabits per second). However, Netflix recommends a minimum of

3.0 Mpbs for HD streams, and 5.0 Mbps is required for Super HD. Furthermore, Netflix will dynamically adjust its resolution level upward or downward to match a customer's bandwidth, choosing automatically between more than 120 different "recipes" of video encoding to find the best match for the consumer's device given the available bandwidth (Ueland 2015). At the bottom end, "a file encoded with a bitrate of 235 kbps . . . would work even on very slow connections, but also only deliver a resolution of 320 by 240 pixels" (Roettgers 2015). Since 2016, there has also been an option within the Netflix app to download shows over wifi to watch later—a feature designed for commuters, users with irregular access to wifi, and those who need to carefully manage their bandwidth.

These bandwidth demands present a problem for many users, given that reliable high-speed connections are not always available outside the major cities. Many populous nations with booming middle classes, including Indonesia, India, and the Philippines, all have sub-3.0 Mbps average speeds for wired connections and much slower and more expensive connections via mobile devices. This effectively makes the full Netflix experience inaccessible to many users in these countries. (This is not just an emerging-world problem, for several rich countries, including Australia and Taiwan, barely meet the 3.0 Mbps average threshold.) While Netflix is now potentially available to users in almost all countries, access to the service depends in practice on both the reach and the capacity of a country's broadband infrastructure, as well as the pricing structures that regulate both internet access and SVOD services. This varies considerably both between countries and within countries, with urban areas typically much better served by telcos than rural areas are. Taking such variables into account, one sees how the global expansion of premium streaming services is unlikely to extend very far beyond the urban middle classes, at least in the medium term. Premium video streaming services are fundamentally different from basic internet protocols, such as email, which have greater flexibility and capacity to cater to mobile-first or mobile-only and low-bandwidth users. The need for credit card payments also restricts the diffusion of subscription services. In other words, there is a spatial as well as an economic logic at work in determining the scale and extent of streaming takeup.

This tension between promised and actual availability can be seen in the case of Cuba. In February 2015, Netflix announced that it would be one of the first American companies to do business there following the Obama administration's removal of Cold War–era trade sanctions against the island state. A Netflix (2015b) press release quoted Reed Hastings as saying that, "We are delighted to finally be able to offer Netflix to the people of Cuba, connecting them with stories they will love from all over the world." However, as Fidel Rodríguez (2016) notes, the barriers to access in Cuba cannot be surmounted by flicking a switch in Los Gatos. Cuba's extremely restricted and slow public internet makes watching Netflix virtually impossible. Furthermore, nobody in Cuba can legally use Netflix, because Cubans do not have credit cards and Netflix does not accept local payment. There is also the issue of pricing: Netflix's Cuba service costs $7.99 a month, but the average Cuban wage is US$17 a month. In other words, Netflix's Cuba service exists only in a virtual or theoretical sense. Infrastructurally and in terms of actual user practices, the service has no meaningful presence.

Sociospatial differences in connectivity *within* a country also shape access to digital services. In many developing countries, while elites may have access to high-speed home internet connections, the majority of the population are either offline or access the internet through mobile devices (often sharing devices among friends and family, with prepaid credit purchased in small amounts) and public cybercafés, kiosks, and shared-password wifi connections. In this context, access to SVOD services is limited not only by geography but also by the nature of internet use "on the ground" and the commercial and social practices around it, which may make it difficult to install apps, use personalization features, and so on. Hence, even when high-speed internet infrastructure exists in a particular city, pricing, practices, and social context may be more important in determining levels of access. This adds another layer to the digital inclusion problem, meaning that any maps of global SVOD usage require multiple overlays to understand not just what cities and regions but also which communities within them will be able to enjoy streaming services as intended.

Politics of Bandwidth

Netflix is aware of these hard and soft infrastructural barriers and what they mean commercially. The company's future is dependent on the takeup of high-speed internet by the global middle classes; in the long term, this is where the growth in subscriber numbers will come from. For this reason, Netflix expends considerable energy lobbying for investment in internet infrastructure, both directly and through various intermediaries.

Netflix's public relations around these issues include some clever initiatives that seek to engage users in infrastructure debates. For example, Netflix has developed a simple internet speed test for users (fast.com), as well as the Netflix ISP Speed Index (ispspeedindex.netflix.com), which ranks countries according to their highest, lowest, and average internet speeds, as detected by Netflix servers. It also ranks ISPs in each country on the same basis. Described by Evan Elkins (2018: 2) as "consumer education projects presented through the altruistic rhetoric of global Internet infrastructure development," these tools encourage scrutiny of ISPs in terms of how well each delivers the Netflix experience. Table 3.1 shows the results for selected countries, taken from the Global ISP Index in late 2016. Clicking on any of these countries will take you to another table, listing the major ISPs in that country and their average download speeds. In Portugal—to give a random example—four ISPs are listed (MEO, Cabovisao, Nos, and Vodafone), with average speeds ranging from 2.85 to 3.57 Mbps.

The ISP Speed Index is now used widely by Netflix subscribers as well as other parties seeking comparative data on internet speeds. Its apparent transparency (it claims to provide a purely technical diagnosis) belies a wider policy by Netflix to name and shame underperforming ISPs—and the internet infrastructure of entire countries—to encourage scrutiny, advocacy, and investment in internet infrastructure. While internet users in many countries would support these principles, we should bear in mind that this is a commercial calculus on Netflix's part. Perceived long-term benefit to digital service providers, achieved through market growth, is the driving force behind these public campaigns.

TABLE 3.1. Average global internet speeds for selected countries according to Netflix

Country	Average Speed (Mbps)
Venezuela	1.11
India	1.83
Jamaica	2.35
Malaysia	2.84
Australia	2.88
Mexico	2.99
Taiwan	3.03
Thailand	3.04
Canada	3.15
Spain	3.15
France	3.23
Indonesia	3.29
Italy	3.30
USA	3.34
Japan	3.45
Germany	3.72
Sweden	3.84
Switzerland	4.11

Data source: Netflix ISP Speed Index, November 2016, selected countries.

There is an institutional dimension to these policy debates. Netflix is a key player in the Washington-based Internet Association ("the voice of the internet economy"), which is active across a range of issues, including intermediary liability and internet freedom, and whose other members include Facebook, Google, Dropbox, eBay, and Spotify.[3] In its home market, Netflix has also been a vocal critic of capped internet plans (Brodkin 2016). Like many other technology companies, Netflix is clearly engaged—for its own commercial reasons—in long-term lobbying for investment and beneficial regulation of tele-communications. Like mining companies that pressure governments to build road and rail connections to their extraction sites, or shipping companies that demand state investment in ports, container infrastructure, and road connections, Netflix is part of a wider internet industry agenda that sometimes blurs the line between public investment and private gain.

This is complex policy terrain, and it is not easy to distinguish corporate self-interest from good public policy. As Davies (2016) suggests, there is a circular logic in many of the discussions about high-speed internet services: video providers, including Netflix, reasonably claim they are adding value for ISPs by creating consumer demand for fast internet access, while ISPs may claim that bandwidth-intensive video services are free-riding on their infrastructure or undercutting their other pay-TV offerings. These tensions reflect the leaky boundaries between different industry sectors, and the consumer demand dynamics that connect them in an internet economy. They also raise difficult ethical questions: Who ultimately pays for the high-speed internet capable of delivering HD video streams? On whose shoulders do the costs fall, and which users are likely to reap the most benefit?

From an infrastructural perspective, the contemporary internet is very much about video. In the United States, Netflix accounts for more than a third of total downstream internet traffic on wired connections (Sandvine 2016a). When combined with YouTube's traffic, this figure jumps to around 50%. Consider Table 3.2, which shows the relative significance of video services as opposed to music and web applications. This thirst for video has implications not only for digital divide debates but also for wider

TABLE 3.2. Top ten peak period applications and their relative bandwidth use—North America, 2016

Netflix	35.15%
YouTube	17.53%
Amazon Video	4.26%
HTTP (other)	4.19%
iTunes	2.91%
Hulu	2.68%
SSL (other)	2.53%
Xbox One Games Download	2.18%
Facebook	1.89%
BitTorrent	1.73%

Data source: Sandvine (2016a).

policy debates about who ultimately benefits from investment in internet infrastructure. For example, it could be argued that—based on overall trends in internet traffic—a disproportionate amount of the value that arises from high-speed internet accrues to providers of video services and their users rather than to those who make do with basic email and web browsing. The Sandvine reports note that this hunger for video is apparent in other parts of the world as well, although it is more often directed toward free services, notably YouTube and BitTorrent, rather than toward subscription services like Netflix.[4] So, while recognizing the apparent disparity, we should be mindful that demand for video is not just a first-world phenomenon, although capacity to deliver on that demand is largely concentrated in rich nations.

ISPs and telcos are naturally wary of high-definition video because of the demands it places on their networks (though these same companies may benefit commercially from increased demand for high-speed and uncapped data plans). Even if we do not trust the motives of these companies—and there are good reasons to be skeptical—there is still a substantive issue here about infrastructure. High-speed internet is a scarce, inequitably distributed resource, subject to struggle and vested interests, rather than a ubiquitous feature of modern life.

Netflix and the Net Neutrality Debate

In the United States, public discussion around these issues often centers on net neutrality, or whether ISPs should be able to charge service providers extra fees for "fast lane" treatment (and, conversely, whether nonpaying services can be slowed down, by design or in terms of relative performance). Netflix has played an important, and somewhat controversial, role in this ongoing debate.

The net neutrality issue came to a head in 2014, when subscribers to some of the major ISPs started to complain about poor streaming quality on Netflix. Netflix reluctantly started paying interconnection fees to Comcast, AT&T, and Verizon to resolve "congestion at the connection point where we transfer content to the ISP" (Hastings 2014)—and thus to improve some subscribers' streaming experience, which within days went from VHS quality to Super HD quality. At the same time, its executives mounted a public campaign to draw attention to the risks of such payments. "Customers pay companies like AT&T, Comcast, and Verizon a monthly fee, and some are even financially penalized if they exceed usage caps," Reed Hastings (ibid.) wrote in *Wired*, "Charging us a separate fee ultimately means consumers pay twice—first for their broadband connection and second through higher-cost or lower-quality Internet services."

At issue here were interconnection fees, or payments by service providers to ISPs for direct connection to their networks—a practice that is technologically distinct from fast-lane payment. Interconnection fees are seen by some industry observers as a routine part of network traffic management (Rayburn 2014); Netflix, however, sees them as a kind of profit-gouging. Interconnection payments may be best understood as an additional layer of soft infrastructure that has evolved over time through negotiations between platforms and ISPs, reflecting the commercial evolution of network traffic management in the United States rather than the inherent properties of the underlying network.

In a response posted online, which is worth reading in its entirety for what it reveals about the politics of video streaming, AT&T CEO Jim Cicconi (2014) offers a rebuttal to Hastings. Cicconi argues that video-on-demand services are "driving bandwidth consumption by consumers to record levels," and ISPs are therefore obliged to "build additional capacity to handle this traffic," which means some costs must be passed on to consumers. The real question, according to Cicconi, is about which consumers should pay for Netflix's traffic:

> If there's a cost of delivering Mr. Hastings's movies at the quality level he desires—and there is—then it should be borne by Netflix and recovered in the price of its service. That's how every other form of commerce works in our country. It's simply not fair for Mr. Hastings to demand that ISPs provide him with zero delivery costs—at the high quality he demands—for free. Nor is it fair that other Internet users, who couldn't care less about Netflix, be forced to subsidize the high costs and stresses its service places on all broadband networks. (Cicconi 2014)

These comments and the net neutrality issue more broadly reflect the blurred line in internet policy between public and private infrastructure, and public and private needs. What are the ethics of bandwidth intensity, and what do they mean for consumers and nonconsumers of video services? How do we allocate the real costs of internet infrastructure?

Granted, this is not a zero-sum game; high-bandwidth uses and users do not have to crowd out other kinds of uses and users. Netflix would point to a range of ways in which it is containing its overall bandwidth demands: by improving its encoding processes, for example, and by building its own content delivery network (discussed later), Open Connect, to take pressure off the public internet. Indeed, Open Connect now carries the vast majority of Netflix video traffic (Florance 2016), rendering moot some of the earlier concerns about net neutrality. From a technical perspective, the ISPs' argument about video services hogging available bandwidth is also simplistic, because there are so many factors that interact to determine end-user speed and experience. The image of congestion, like a traffic jam on a highway, is misleading. Finally, we should not forget the unique conflicts of interest and regulatory failures that lie behind this debate in the United States, where ISPs are also cable TV providers. The potential for abuse of market power is quite real, as are the risks of introducing further conflicts of interest into the system.

Nevertheless, there is something disquieting about Hastings's insistence that "broadband is not a finite resource" and that "network limitations are largely the result of [ISPs'] business decisions to not keep pace with subscriber demand" (Hastings 2014). Perhaps it is the vision of digital infinitude that underlies this statement—the idea of more and more devices requiring ever-higher definition and ever-faster connections—or perhaps it is the flat refusal to countenance any material consequences to the race for high-definition video. This is a vision that can never scale meaningfully in a global sense. It is grounded in a first-world idea of the internet, premised on an assumption of unbounded capacity. It does not ring true with how the internet is experienced in many countries, including my own (Australia), where high-speed internet access is still relatively scarce because of our aging copper-wire phone network.

While not begrudging Netflix's position here—which is commercially sensible within the context of North American net neutrality debates—as critical media scholars we need to interrogate the implicit geography of inclusion and exclusion that is embedded within this vision of the internet's future. Such are the complexities that emerge when we take an infrastructural view. Having considered some of the constraints that shape the geography of video-on-demand over the public internet, let us now move on to consider some additional layers of infrastructure that sit within Netflix's internal systems architecture.

Clouds and CDNs

As we move further into the internet age, the evergreen Marxian imperative to follow the money and uncover the truth has been augmented with a second imperative: to follow the data. Current media studies research on data centers and the geography of networks challenges us to think critically and spatially about where data resides, how and where it travels, and who controls these movements (Starosielski 2015; Holt and Vonderau 2015; Rossiter 2016).

Consider the following account by a *New York Times* journalist, written in 2009, about the geography of their own personal data:

> I have photos on Flickr (which is owned by Yahoo, so they reside in a Yahoo data center, probably the one in Wenatchee, Wash.); the Wikipedia entry about me dwells on a database in Tampa, Fla.; the video on YouTube of a talk I delivered at Google's headquarters might dwell in any one of Google's data centers, from The Dalles in Oregon to Lenoir, N.C.; my LinkedIn profile most likely sits in an Equinix-run data center in Elk Grove Village, Ill.; and my blog lives at Modwest's headquarters in Missoula, Mont. (Vanderbilt, cited in Sandvig 2015, 90)

Here we have a very complex yet manageable scenario because both the user and the servers are located in the same country. In the case of Netflix, a global user community is accessing a U.S.-based service, making things considerably more complex in terms of the geography of information sent and received. What actually happens, and where does the data go, when a user in Manila or Manchester fires up the Netflix app?

There are two layers of infrastructure that we must understand before we can follow the data to answer such questions. The first controls the Netflix user interface, customer data, recommendation algorithms, personalization, and other elements of the platform (what you see *before* you hit Play). The second is Netflix's content delivery network, called Open Connect, which stores video and audio content in servers located close to end users (this controls what happens *after* you hit Play). Let us start with the preplay data. In the past, Netflix used to serve everything from its own data centers in the United States. After a major crash in 2008, it started progressively moving to the public cloud, where its service could be more easily scaled. Netflix now uses Amazon Web Services (AWS)—a profitable arm of Jeff Bezos's Amazon e-commerce empire—exclusively for this purpose; hence understanding the geography of Netflix data requires that we know something about AWS.

AWS has servers all over the world. As of 2016, the three AWS "regions" used by Netflix were located in Northern Virginia, Oregon, and Ireland.[5] Data are served from the AWS region closest to the customer: subscribers in Europe, the Middle East, and Africa are served from Ireland; subscribers in the Asia-Pacific region are served from Oregon; Latin American subscribers are served from Northern Virginia; and U.S. subscribers are served either from Oregon or Northern Virginia (Madappa et al. 2016). In the case of a crash, the system is also designed in such a way that customer data can be fetched from any of the other AWS servers. Not every request is actually processed in real time by these servers. Netflix precomputes some calculations (especially recommendations) to cut down on processing time and cost (Amatriain and Basilico 2013).

Now consider what happens after you push Play. Video providers have to think carefully about their network organization, given the massive amounts of bandwidth involved. Things work best when they *pre-position* content as close as possible to the end user, so that data delivery is as cheap and fast as possible. This is the job of the CDN, a network of distributed servers that reduce latency in video transmission by caching popular video content in local

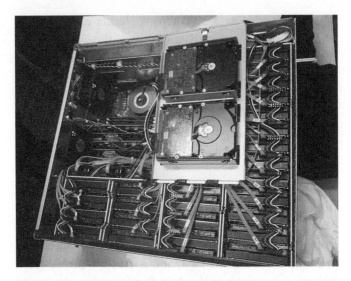

Figure 3.2. A Netflix Open Connect server in 2012. Image by Andrew Fresh (CC-BY2.0 license).

servers closer to the end user, thus reducing load times and buffering. Many major web services use a commercial CDN, such as Akamai or Amazon CloudFront, which have global networks of servers designed for this purpose.

Netflix is a special case. It used to contract with third-party network companies for CDN services, but in 2011 it decided to start building its own CDN to bring this infrastructure under its direct control. This in-house CDN, called Open Connect, has since become a vital part of the Netflix service, delivering over 125 million hours of viewing per day.

The system relies on thousands of boxes such as the one pictured in Figure 3.2. Netflix provides participating ISPs[6] with these Open Connect server boxes, each of which contains a full Netflix library—roughly four years of HD video. ISPs place the Open Connect servers in their network so that users streaming Netflix can connect directly to the box rather than to faraway servers over the public internet. Each day, during off-peak hours, the Open Connect servers are refreshed with a copy of every Netflix title so that they have an up-to-date catalog of content. Netflix uses predictive modeling to estimate the demand for certain kinds of titles in certain places, and it tries to ensure there is sufficient network capacity within local Open Connect servers to serve all those streams effectively. As Ken Florance, Netflix's vice president of content delivery, explains,

> We now have Open Connect Appliances in close to 1,000 separate locations around the world. In big cities like New York, Paris, London, Hong Kong, and Tokyo, as well as more remote locations—as far north as Greenland and Tromsø, Norway and as far south as Puerto Montt, Chile, and Hobart, Tasmania. ISPs have even placed OCAs [Open Connect Appliances] in Macapá and Manaus in the Amazon rainforest—on every continent except Antarctica and on many islands such as Jamaica, Malta, Guam, and Okinawa. This means that most of our members are getting their Netflix audio and video bits from a server that's either inside of, or directly connected to, their ISP's network within their local region. (Florance 2016)

Open Connect is a massive, paradoxical thing—a private network built on top of the public internet. As such, it is beneficial for ISPs, Netflix users, and Netflix nonusers alike because it relieves pressure on internet infrastructure. But it also represents, from a certain perspective, a privatization of the idea of the public internet—a purpose-built fast lane to the consumer, reserved for Netflix traffic. Open Connect allows a quality of service that gives Netflix a significant competitive advantage over other streaming sites, few of which could afford such elaborate infrastructure.

We tend to think of the internet, with its egalitarian routing procedures, as "a silky web in which every place is equally accessible to every other place" (Blum 2012, 6), but at the infrastructural level this idea does not hold water. The complex network engineering strategies used by major providers—including peering or interconnection payments, CDNs, and so on—remind us that internet infrastructure is not neutral: there are many ways in which certain kinds of traffic can be made to move faster and to connect better than other traffic. Indeed, the rise of CDNs presents regulatory problems that we are only now beginning to understand. As Palacin et al. (2013) note, while the net neutrality debate focused on the provision of fast lanes for certain kinds of traffic, it has been remarkably silent about CDNs and other network arrangements that similarly work to prioritize some uses and users over others:

> CDNs are not being considered as violating network neutrality principles, although they offer "faster lanes" for those content providers who can afford it, possibly also leading to a two-class (or more) Internet. In this context, one can argue that CDNs are not degrading the rest of the traffic, but how can a long tail video website compete against a "hyper-giant" whose content is distributed using high speed connections? (Palacin et al. 2013, 323)

Current debates about internet policy often overlook the "transparent intermediaries" (Braun 2015), such as

CDNs, that play a vital role in digital media distribution. These constitute an additional, opaque layer of mediation, because of their private nature. Netflix, in its defense, can rightly point to its considerable investment in Open Connect as representing its contribution to carrying the costs of network capacity, as Cicconi demanded.

Looking closely at the Open Connect structure also reveals some interesting things about the priorities of Netflix when it comes to their international markets. In a technical paper published shortly after the Netflix global launch, a team of computer scientists (Boettger et al. 2016) mapped the Open Connect network to determine which areas had the most capacity and how traffic flowed across the network. Their analysis points to some strange quirks in the Open Connect network—for example, the tiny Micronesian island of Guam is well served by Open Connect servers, presumably because of its U.S. military bases, while the "server deployment in Africa is very limited, representing a tiny market for Netflix" (Boettger et al. 2016, 9). Reading across all these measurements, the authors conclude that "the latest expansion in 130 countries announced by Netflix in January 2016 was only a virtual expansion" and that "many countries where Netflix is officially available still remain without any Netflix infrastructure" (10). This attempt to follow the infrastructure reveals a rather different story about Netflix's globalization than the one suggested in the company's public relations. It gives us a sense of the unevenness of Netflix's presence around the world, and its finite capacity to deliver on its promise to be a global television network (recall Hastings's words at CES: "No more waiting . . . no more frustration. Just Netflix"). The infrastructural footprint of Netflix, while formidable, is certainly not everywhere, and is tailored around the company's idea of what the most important global markets might be and where they are located. The geography of infrastructure is, once again, inextricable from the geography of capital.

The Long View

This chapter has traced the rough geography of Netflix's hard and soft infrastructure to better understand the degrees to which Netflix can and cannot be considered a global media service. Evidently, describing the geography of internet television is no easy thing. There are many different layers of infrastructure here, forming a series of nested maps rather than one single map. As we have seen, the base-level constraint is the availability or nonavailability of high-speed internet service sufficient for video streaming (which is largely restricted to major cities, and usually to elites within those cities). On top of this base layer, there are many layers of soft infrastructure that interact to further shape the actual availability and experience of streaming, including data pricing, ISP network policies, payment systems, and others. There is also the internal network architecture of Netflix's CDN, which works to effectively prioritize certain regions over others when it comes to speed and user experience.

These layers of infrastructural inclusion and exclusion interact to shape the actual experience of internet television. It does not ultimately matter to a user whether their video will not load because of the slow speed of a DSL connection, congestion at the interconnection point, the absence of a local CDN, or because they have run out of data on their mobile plan. The result is essentially the same. As scholars, we should be mindful of the infrastructural inequalities that underlie these scenarios. At the same time, our analyses must drill down into the finer details of how various layers of infrastructure are interacting in the background to produce such an effect.

A second insight arising from the discussion so far relates to how we think about Netflix itself—its infrastructural ontology, if you will. Netflix is not a unitary thing but a complex and dynamic metasystem made up of hundreds of different software processes that relies on *hard* and *soft* technical infrastructures, *open* and *closed* knowledge systems, and *public* and *private* investment. These apparent contradictions are built into the platform, and their echoes linger in the ongoing debates that Netflix has become embroiled in, such as net neutrality. In some respects, Netflix can also be seen as an infrastructure builder. It has also developed a content delivery network of its own, as well as various software tools and processes, some of which it releases back into the world as open-source projects (McEntree 2010). It is also involved in policy and advocacy around internet infrastructure generally, especially in the United States.

Peering under the hood—or, in Netflix's case, into the network—is a useful exercise. I have suggested that the tactic of looking through the object and out to the infrastructure is essentially an extension of Marxian critical techniques—the imperative to trace the circuitous paths of the commodity, revealing the materiality of industrial production that forms its constitutive secret (from the sweat of the worker to the glittering commodity in the display case). In the case of video-on-demand, this also involves tracing the path of data through networks to understand where our digital entertainment comes from, what territory it moves through on its way to us, and how its passage is materially and politically governed—in Braun's words, the "kinks, epicycles, and roundabouts found in a distribution route" (Braun 2015, 9).

Infrastructural thinking is above all a mode of theorizing. It allows us to see (indeed, demands that we see) media systems from unusual perspectives. It can help us to think in completely different terms about those objects. In a lucid essay, media scholar and anthropologist Brian Larkin (2013, 329–330), following Paul Edwards, uses the computer as an example to show how infrastructural thinking can productively dissolve the boundaries around a given object so that the "simple linear relation of foundation to visible object turns out to be recursive and dispersed":

> Take, for example, the computer I have used to write this article. What is its infrastructure? Electricity may be the most obvious substratum that allows the computer to operate. But, as Edwards (1998) notes, although electricity is the infrastructure of the computer, the computer is the infrastructure of electricity supply, as the entire transmission industry is regulated by computers. Electricity, in turn, has other infrastructures [including] oil production[,] . . . financial mechanisms . . . that allow electricity to be sold on an open market, or the labor networks necessary to produce and transmit power. (Larkin 2013, 329)

When we start thinking infrastructurally about media, the distinction between foreground and background, and between object and context, can start to blur. Such a move brings new kinds of relations into view. What if we were to apply this kind of thinking to Netflix? Instead of looking *behind* the object and into the constitutive back end, what might happen if we looked *out from* Netflix at the various other infrastructures to which it is in some way connected?

Following Edwards and Larkin, an infrastructural view of Netflix would include not only the systems on which it relies (electricity, telecommunications, internet governance) but also more obscure technical infrastructures such as standardized text and character inputs, video and audio encoding standards, metadata formatting, and user interface design standards, as well as international banking transfers, credit card systems, personal credit rating providers, and payment processing systems.

From this perspective, the story of Netflix has as much to do with *longer-term, larger-scale social and technical transformations* that lie well outside the boundaries of media studies. These include the history of electrification and lighting, which provide the basic conditions for domestic leisure as we know it today; heating, air-conditioning, and other forms of climate control, as well as modern plumbing and sewerage, which have helped to create a modern idea of home as a space of comfort protected from the natural elements; modern architectural forms premised on separation of private and public space, with family rooms, TV rooms, and bedrooms in which Netflix can be enjoyed; and social practices of family rearing that welcome technology into the home. The advantage of this explanation for the uneven availability and uptake of Netflix globally is that it does not seek to essentialize the leisure practices associated with a particular urban condition into a general diagnosis about needs and wants. Nor does it presume that people outside the internet television market for whatever reason do not necessarily wish to have something like Netflix in the future—at their price point, in their language, and in a format appropriate to their needs. Instead, it invites us to see technology use in relation to particular trajectories of infrastructural development that come together in particular places at particular times. It helps to explain the cultural specificity of Netflix—which is clearly bound up with a very particular kind of idea about leisure that, as we have seen throughout this book, does not always travel well internationally—in more structural terms.

To the extent that Netflix is a global media service, then, its relative popularity may well be a function of these urbanization processes that produce the basic conditions for its enjoyment—domestic space, disposable income, connectivity, and consumer society—as much as advances in video compression and consumer hardware. Here, paradoxically, we see the strongest continuities between the new internet television services and older forms of broadcast television. Just as Raymond Williams described television in the 1970s as a medium of "mobile privatisation"—"one of the characteristic 'machines for the home'" (Williams 1974, 4)—the account of Netflix's infrastructural geography offered in this chapter reaffirms this fundamentally important observation about the nature of television *as a domestic technology*. Netflix's fundamental vision of entertainment is a personalized experience built around the individual consumer/family unit, equipped with their own credit card and data profiles, to be enjoyed in private spaces. To the extent that these infrastructural conditions

exist in different places, we can expect there to be some interest in Netflix in those markets. For the rest of the world,
it is likely that SVOD services may continue to hover somewhere between novelty on the one hand and irrelevance on
the other.

4

Making Global Markets

To put the internationalization of Netflix in historical perspective, let us go back in time to August 1981, when a new and disruptive cable channel called MTV: Music Television was launched in the United States. MTV—which began as a joint venture between Warner Communications and American Express and was purchased by Viacom in 1985—was the first channel to show music videos 24 hours a day. Massively popular with American teenagers, MTV was a pop-culture phenomenon in its day and is often credited with inventing (or at least popularizing) the music video format. Within a few years of its launch, MTV was available in almost all U.S. cable households. Soon, MTV executives started hatching plans to take the channel global, starting with the newly liberalized media markets of Europe, Latin America, and Asia. Their aim was to build an international advertising market and youth culture around American pop music, with the music video at its center: "One world, one music" was the company's slogan at the time. "Our goal is to be in every home in the world," stated Tom Freston, president of MTV Networks (Banks 1997, 44).

Figure 4.1. Japanese comedian Ryota Yamazato and Reed Hastings at the launch event for Netflix Japan at the SoftBank Ginza store in Tokyo, September 2, 2015. Photo by Ken Ishii/Stringer.

MTV Europe was launched in 1987, and four years later Viacom launched MTV Asia, reaching 42 countries by satellite. These channels were strongly based on the American MTV template and tended to play the same mix of English-language pop videos. The assumption was that MTV's powerful brand would be enough to hook global audiences and that it did not need to customize its programming to suit local tastes. The vast majority of video clips featured American and British artists (Banks 1997, 48). MTV executives justified this approach by stating that "music transcends culture" (Hanke 1998, 223). Advertisers such as Coca-Cola and Levi-Strauss, who were backing MTV's international operations, were of the same opinion, believing that rock'n'roll was an "international language" (ibid.).

This one-channel-for-all approach was a failure. Faced with mediocre ratings, MTV learned that audiences outside the United States were generally less interested in Michael Jackson and Madonna than they were in their own stars, styles, and music genres—from mandopop to merengue. Apparently, not everybody wanted MTV, at least not in the same way. MTV executives soon decided they needed to change their approach, and they began to provide more locally relevant programming. Across Asia, Rupert Murdoch's Channel [V]—MTV's archrival[1]—had been tailoring its programming to local tastes with some success, so MTV followed suit. They increased the proportion of videos by local artists, hired multilingual presenters, and introduced customized blocks of programming specific to each territory. Eventually, MTV Asia morphed into MTV Korea, MTV India, MTV Japan, MTV Taiwan, and so on. By the early years of the new millennium, the company's strategy for global markets had evolved. MTV executives now described their network as "a global brand that thinks and acts locally" (Santana 2003). MTV Networks CEO Bill Roedy spoke proudly of the differences between the various MTV channels in Asia:

> MTV India is very colorful, self-effacing, full of humor, a lot of street culture. China's [MTV] is about family values, nurturing, a lot of love songs. In Indonesia, with our largest Islamic population, there's a call to prayer five times a day on the channel. . . . Japan's very techie, a lot of wireless product. (Roedy in Gunther 2004, 116)

As these comments suggest, the MTV strategy had changed from an "American export" model to a network of national channels showcasing diverse music styles and locally produced series and skits. MTV executives had learned two important lessons during this period. Localization matters in television markets, and the global will not simply displace the local.

The story of MTV's misadventures raises issues that are directly relevant to Netflix's internationalization some twenty years later. In both cases, we see the launch of a disruptive American television service, the attempted export of this service to global markets, uneven uptake, cultural blowback, and then a commitment to localization and local content production. MTV and Netflix are quite different—one is a cable/satellite network, the other a subscription streaming aggregator and studio—so we have to be careful about comparing the two. Nonetheless, there is a common pattern here: initial overconfidence, followed by a dawning realization that what works at home does not always work abroad. This chapter will consider some of the challenges Netflix has faced in the major Asian markets—China, Japan, and India—where the limits of U.S. media power are plainly visible.

Global Television, Local Markets

The history of transnational television is full of these frictions between the global and the local. In his influential work on pan-European satellite television, the sociologist Jean Chalaby (2005, 2009) offers a detailed analysis of the localization process for U.S.-based channels, including Discovery, National Geographic, and Bloomberg, which entered Europe during the 1980s and 1990s. Chalaby observes that, at the start of the transnational satellite boom in the 1980s, these channels were initially

> unsure how to transmit across boundaries and were at first oblivious to local culture and market conditions. They had overestimated audience appetite for foreign programming and launched general entertainment channels in direct competition with established national broadcasters. To remedy this, they have progressively focused on niche markets and begun adapting their feed to local tastes. In the mid-1990s, the emergence of networks demonstrated how corporate players had acquired a much better understanding of the relationship between the local and the global, and learned how to *articulate the two polarities while benefiting from both*. (Chalaby 2005, 62, my italics)

Chalaby identifies a spectrum of strategies that channels use to localize their offerings, ranging from superficial repurposing of existing content to a deeper, and more expensive, localization. These options include, in order of increasing complexity, subtitling, dubbing, or adding live voiceover; splitting the original feed to create local windows for additional programming; and creating "a network of local channels around a core broadcasting philosophy," the most expensive and complicated option, which involves considerable investment in original programming (Chalaby 2005, 56).[2]

Chalaby and other scholars who have studied localization in television industries—including Moran (1998, 2009), Straubhaar (2007), and Kraidy (2008)—describe in their work a process of cultural learning through commercial experimentation. By the 1990s, television executives had become cosmopolitans, learning the lesson of cultural difference through trial and error. They understood that international markets do not simply exist, waiting mutely for great content, but must be *made*—which is to say that they must be primed, cultivated, and maintained. Audiences' preference for locally produced content cannot easily be overcome, so it must therefore be respected. These findings lead Chalaby to conclude that the business of transnational television, at least in Europe, is all about the challenge of localization. Serious broadcasters understand that global audiences have distinctive tastes, preferences, and expectations. Local expertise—including staff with a deep understanding of a country's media landscape—is necessary for success. Tailored strategies and programming are needed for each market. The conceptual correlate of this is that the logic of capital does not necessarily lead to standardization, as per the media-imperialist critique of multinational media corporations, but can also lead to increased *difference*.

This literature on transnational television and localization offers important insights for understanding Netflix's situation as it seeks to compete in more than 190

Figure 4.2. Reed Hastings at the launch event in Bogota, September 9, 2011. Photo by Getty Images.

Figure 4.3. Marketing campaign for *House of Cards* season 5 in Warsaw ("It's time for me to take the helm"), June 2017. Photo by Grand Warszawski/Shutterstock.com.

Figure 4.4. In-store display promoting Netflix's tieup with the French telco SFR—seen in Lyon, July 1, 2017. Photo by the author.

Figure 4.5. Promotion for the Spanish-language Netflix original series *Las Chicas del Cable* in the Puerta del Sol, Madrid, April 2017. Photo by Curtis Simmons/Flickr (CC BY NC 2.0 license).

international markets simultaneously (albeit with different levels of commitment to those markets). As we have seen, the tendency among multinational media companies is to begin the internationalization process with an undifferentiated image of the global market, seen as a flat space waiting for innovative content (recall Reed Hastings's remarks at CES—"no more waiting!"). Over time, this undifferentiated image of the global market is replaced with a more complex understanding of national *markets*, seen as discrete containers characterized by diverse tastes, income levels, languages, genre preferences, willingness to pay, and other factors. The challenge of global media is to adapt to these diverse conditions. Success, then, is not just about pushing out great content to the world but also about understanding and negotiating cultural differences.[3] While Netflix initially struggled with this hard truth, it has fully absorbed this lesson over the course of its internationalization.

Long-Distance Localization

One of the distinctive features of internet-distributed television services is that they can enter into and compete in a large number of international markets without extensive in-country infrastructure. For earlier transnational television channels and film studios setting up in foreign markets, market entry used to require "boots on the ground"—a local office, sales and customer service teams, local agents to advise on strategy, and partnerships with producers, brands, advertisers, and telcos (Thompson 1985; Donoghue 2017). This was seen as the necessary work of media globalization. As Michael Aragon, a senior executive at Sony Network Entertainment, said in 2011 when asked about the challenge of securing local content for international operations: "You can't just run this out of L.A. . . . That was an assumption we quickly blew out of the water" (Aragon cited in Wallenstein 2011). Netflix confounded this logic by coordinating most of its international expansion from its base in California. When entering the key European market of France, Netflix employed only "a stripped down Paris-based team of three people" (Goodfellow 2015). In Canada, its oldest international market and still one of its most important territories, Netflix never opened a local office until it began producing in earnest there after 2017. At the time of the Australian Netflix launch, the only local staffer on the ground who could answer press inquiries was a freelance publicist.

Netflix now has offices in many countries—including the Netherlands, Singapore, Taiwan, Japan, Mexico, Brazil, England, and India—but it centralizes its operational activities in only a few locations. Most customers are billed through Singapore, Amsterdam, or Los Gatos, for instance.[4] Programming and marketing for many countries is managed out of the Beverly Hills office. Wherever possible, Netflix prefers to invest in its technical infrastructures rather than spending money on foreign offices and staff. It mines its greatest asset—its vast stores of customer data—to identify trends that help it understand what kind of programming current and potential Netflix subscribers in international markets might want.[5]

Netflix's approach to localization was a source of derision for industry observers from the old world of broadcasting. John Medeiros of the Cable & Satellite Broadcasting Association of Asia mocked the approach:

"They [Netflix] sit in Silicon Valley, open the gusher, and the sweet crude flows all through the global network, just like that? . . . Maybe that happens (for a while) in a technical sense, but in a commercial sense, you still need access to customers and to their payments, and the governments that you're flipping off might have something to say about that" (Medeiros cited in Frater 2016). Yet the structure has been effective in the sense that it enables Netflix to run most of its most labor-intensive activities from a few key global hubs, achieving economies of scale. The logic of long-distance localization extends the spatial patterns begun with satellite broadcasting in the 1980s, which precipitated this separation of a broadcaster's service zone from its operational infrastructure.

The Unavoidable Labor of Localization

While in-country overhead can be minimized, there is still an enormous amount of work that needs to be done on the Netflix platform to make the service relevant and legally compliant in international markets. For example, content, interfaces, and artwork must be translated; content must be classified to conform to local laws; country-specific catalogs must be programmed and maintained; catalog categories must be tweaked for each country to highlight local content; payment systems must be adapted so that people can pay with local debit cards and through their mobile carriers; billing systems must be flexible enough to collect national and state-based sales taxes and repay them to tax authorities; and age-verification and PIN protection mechanisms must be introduced to comply with national laws. Taking this into consideration, over-the-top distribution does not erase the *need* for localization; it just

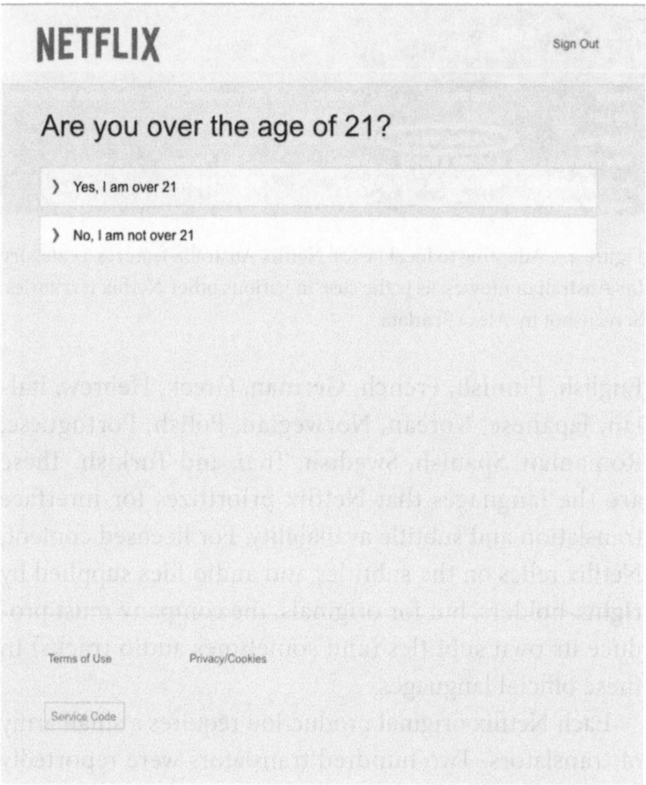

Figure 4.6. Adapting to local laws: Netflix Singapore features an age-verification system for R21+ content. Screenshot by Alexa Scarlata.

reorganizes this labor across new spaces. As an example, consider how Netflix manages the essential business of translation. Netflix's translation efforts are organized around its priority list of "official languages." Initially this was a small list, including English, Spanish, and Portuguese, in the early years when Netflix was available only in the Americas, but over time it has grown to more than 20 languages—Arabic, Chinese, Danish, Dutch,

Figure 4.7. Adapting to local tastes: Netflix Australia features a category for Australian movies, as is the case in various other Netflix territories. Screenshot by Alexa Scarlata.

English, Finnish, French, German, Greek, Hebrew, Italian, Japanese, Korean, Norwegian, Polish, Portuguese, Romanian, Spanish, Swedish, Thai, and Turkish. These are the languages that Netflix prioritizes for interface translation and subtitle availability. For licensed content, Netflix relies on the subtitles and audio files supplied by rights-holders, but for originals, the company must produce its own subtitles (and sometimes audio tracks) in these official languages.

Each Netflix original production requires a small army of translators. Two hundred translators were reportedly needed for subtitling one season of the Netflix original talk show *Chelsea* (Roettgers 2017). Netflix cannot do all this work in-house, so it uses a large network of freelancers and also subcontracts with firms in the United States, Poland, Spain, Australia, Mexico, England, and Sweden.[6] These companies form part of the GILT (globalization, internationalization, localization, and translation) sector, a relatively new industry that has been growing since the 1990s and now employs thousands of localization experts across the world who specialize in making websites and apps work effectively for an international user base.[7] These are the unsung foot soldiers of digital media globalization, the people who know how to smooth out the frictions of culture and make services like Netflix work everywhere.

The number of variables to consider when translating for Netflix is enormous. Some licensed content has subtitles hardcoded into the video file. Older viewing devices are not always able to process non-Latin alphabets. Fonts must be carefully chosen, mindful of local connotations and visual aesthetics in different countries.[8] There are also issues on the supply side. Even when subtitles exist for licensed content, they are not necessarily made available in all Netflix territories (studios withhold the full range to discourage out-of-region access). This has prompted a workaround that is known to many Netflix users and seems to be tacitly endorsed by the company: one can trigger a secret menu in the desktop browser interface, then upload a subtitle file in DXFP format, and this will play on top of your chosen video.[9]

Interface translation has its own specific requirements. Arabic reads right to left, while Latin languages read left to right. The length and spacing of words changes from language to language, and thus the interface design may need to be adjusted to retain the standard look and feel. This requires some complex programming to build templates that can then be automatically filled with the correct text content. In addition to menu options, the display of content titles, ratings, and metadata may or may not change according to location. These illustrative examples from an evolving system give us a sense of the scale of backstage work required for Netflix to run effectively in multiple markets.

At the time of writing, Netflix's translation is still done by humans, though the company is keen to standardize this work as much as possible. In 2017, the company introduced a new in-house system called Hermes, described as "the first online subtitling and translation test and indexing system by a major content creator" (Fetner and Sheehan 2017). Until it was shut down in 2018, apparently due to an overwhelming response from applicants, Hermes was Netflix's automated system for registration, testing, and accreditation of translators. Potential employees would create a profile on Hermes and then take a test to assess their skill in translating idiomatic English expressions ("made a killing," "hit the road") into their own language. Test results determined the amount and kind of work Netflix would then offer the translators through the platform. Hermes was just one of the many ways that Netflix, in classic Silicon Valley style, has been trying to engineer efficiency into the "cultural" business of translation.

How do we critically evaluate Netflix's translation efforts? There are at least two ways of thinking about this. On the one hand, we could criticize the unavailability of minor-language translated content, noting that Netflix, rather than being a truly global service, tends to cater exclusively to the major language groups—only 20-odd languages out of more than 6,500. From this point of view, Netflix falls short of its ambition to "delight viewers in 'their' language" (Netflix 2017a), because it has not invested enough in minor-language interface translation, subbing, and dubbing. A more charitable assessment would be that Netflix is now translating more content into more languages than almost any other media service has done or could do. Millions of hours of content are translated annually, and savvy users can add to the stock of subtitles by uploading their own. Seen from this perspective, Netflix may well be the most multilingual television service that has ever existed.

How successful have Netflix's localization efforts been? The answer to this question depends on where you ask it. In the sections that follow, we will consider how Netflix has fared in three important Asian markets: India, Japan, and China. I gratefully acknowledge the work of Thomas Baudinette, Wilfred Wang, and Ishita Tiwary, who assisted me with this research.

India

India, with a vast English-speaking middle class, is a prized market for multinationals. Internet penetration and smartphone use are rising rapidly among an overall population of 1.3 billion people. Satellite and cable systems are well established, with hundreds of available channels. Yet India's demography, regulations, and infrastructure pose many challenges for foreign media companies. Higher-speed internet connections are still largely restricted to urban elites, and their typical bandwidth speeds of 1–2 Mbps are too slow to use video streaming services reliably. Meanwhile, data caps of 40–50 gigabytes per month mean that even those subscribers with high-speed home connections are often wary of using up their precious gigabytes on data-hungry video applications.

India has no fewer than 14 official languages, including Bengali, Gujarati, and Punjabi, as well as the widely spoken Hindi. There is no single language that is understood by all citizens across this vast nation. Proficiency in English, the default language of business, advertising, and elite communication, is limited to around a third of the population. What's more, India's film industry, Bollywood, is an unassailable force: it is by far the most popular form of moving-image entertainment, with Hollywood coming in a distant second. India has its own thriving star system, its own storytelling traditions, its own dream factory. There is no obvious *need* for foreign content, even though the market is certainly big enough to support a range of styles.

In India, Netflix has been licensing Hindi, Gujarati, and Punjabi shows, but at the time of writing its Indian content collection is still seen as poor compared to what is available on other services. Instead, Netflix is reliant on the pulling power of its original content among English-speaking elites. As a Netflix executive told the *Times of India* shortly after the local launch, "We are after a small base of English-speaking people, who travel abroad, are wealthy and want to watch the latest shows that are being launched in the US" (Sarkari 2016).

Netflix got off to a late start in India, arriving as part of the global switch-on in January 2016. By this time, the streaming market was already crowded. Rupert Murdoch's Star TV had its own streaming service, Hotstar. Other popular services included Eros Now (from the Bollywood studio Eros), Sony LIV, and DittoTV (from the cable and satellite provider Zee TV). There were also smaller services, such as Spuul (specializing in Hindi, Punjabi, Tamil, Malayalam, Telugu, and regional-language content), Hooq (with a big catalog of classic Indian TV shows), and Muvizz (specializing in independent movies). Amazon Prime Video was another competitor and was starting to invest significantly in local content. Amazon premiered its first Indian original series, the cricket-themed drama *Inside Edge*, in 2017, along with a series of stand-up performances from popular Indian comics. Hotstar had already released a number of original productions, including romances, talk shows, and several comedy series. For its part, Netflix responded by commissioning several high-profile original series, including *Selection Day*, described as a "story of cricket and corruption" and based on a novel by Booker Prize winner Aravind Adiga; *Sacred Games*, an organized-crime drama also based on a best-selling novel; and the romantic comedy movie *Love per Square Foot*.

Censorship is a delicate issue in India. Movies must receive a certificate from the Central Board of Certification. Explicit sex and violence is typically censored, and Bollywood movies steer clear of this material entirely. On television, offensive terms (including common words like sex, hell, boob, and beef) are muted or cut from the subtitles, while sex scenes and nudity are taboo. A number of U.S. cable channels operating in India, including Comedy Central and Fashion TV, have recently been temporarily taken off the air by Indian regulators for violating these standards.

Digital media services are not regulated under the same law and thus occupy a grey area. Global TVOD services—including Google's Play Store and Apple's iTunes—have erred on the side of caution and only sell films that have received a CBC certificate (and thus only sell the censored cinema version if multiple versions exist). Apple Music blocks adult-rated music from its Indian service completely. In contrast, Netflix has stated that it will not be cutting its original content for the Indian market. Vice President Chris Jaffe went out of his way in January 2016 to reassure Indian subscribers that "nothing on the service is censored at this point" (Jaffe in Avari 2016). This signals an appeal to cosmopolitan Indian subscribers, who prefer uncensored content, despite the regulatory risk inherent in this approach.

In January 2017, a Netflix India employee, acting in an unofficial capacity, took to Reddit to promote the service to users.[10] The staffer then stuck around to field a series of questions from other users in what became an unofficial AMA (Ask Me Anything) session about Netflix's Indian service. The resulting conversation, while not necessarily representing the official Netflix line, provides an interesting insight into how Netflix is seen by tech-savvy middle-class Indians. Many users raised the issue of catalog discrepancies, with the Netflix staffer fielding numerous questions about missing titles. ("*Interstellar* is available in Cook Islands and Kiribati but not India. Can somebody explain plis? [*sic*]") The feedback was not all negative. A number of users wrote that they loved the content and the Netflix experience, especially the lack of censorship. One user wrote, "Thank you . . . Netflix, our Lord and Savior from censored butchered content."

Throughout the discussion, the issue of price kept rearing its head as a key source of irritation. User after user criticized Netflix for being so expensive compared to other digital media. "You have to 'Indianise' the price," wrote one user, "For example, Apple Music is R120 per month here as opposed to $10 in US."[11] As these comments suggest, pricing is a major challenge for Netflix—which maintains a more or less uniform global price point across all countries, with a few exceptions—especially in low-income markets. In India, the monthly cost for the Netflix basic package is presently 500 rupees (US$7.50) per month, in line with its price in most other countries. However, the Indian streaming services are much cheaper than this. Eros Now charges only 49 rupees (US$0.70) per month—a tenth of the Netflix price. Hotstar is free

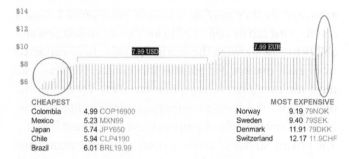

Figure 4.8. Global pricing for the Netflix basic plan, converted to U.S. dollars. Data source: UNOGS.com, September 18, 2017.

with advertisements, or 199 rupees per month for the premium SVOD service with new-release U.S. content and live cricket matches (this price is discounted to 133 rupees per month for upfront payments). Amazon Prime Video is also cheap, at 499 rupees per year (around 42 rupees per month).[12] This massive price gap places Netflix in a different market from its local competitors, making it strictly a premium service for Indian elites, expats, and well-to-do TV buffs. In a country where the lower classes do not have domestic internet access and share content through USB sticks and portable hard drives, paying 500 rupees per month for streaming media is a commitment not to be taken lightly—and not easily available to those without credit cards in any case.

International price setting is one of the most fascinating aspects of the global media economy—a seemingly technical process that actually reflects a whole set of embedded ideas about consumers, money, and value. Digital media services operating transnationally can choose to offer a uniform global price, eliminating the possibility of arbitrage, or they may set a unique price point for each country that can be calibrated to local users' willingness and ability to pay, allowing the maximum value to be extracted from each market. Faced with this choice, Netflix—like most multinational subscription services—chose the former option. The obvious consequence of this choice is that Netflix is unaffordable for all but the most affluent Indians. The pricing strategy "makes it more expensive than many local competitors, and leaves piracy an attractive option" (Frater 2016). This, it seems, is a price Netflix is willing to pay to remain a premium product.

Japan

For a different but equally instructive case, we can look to Japan—a wealthy, media-saturated market that has pioneered many forms of digital distribution. Netflix's Japanese service launched in September 2015 as part of the company's limited Asia-Pacific expansion, comprising Japan, Australia, and New Zealand. Initial launch pricing was 650 yen (US$5.50) for the basic package, rising to 1450 yen for the Premium multiscreen HD package. This put Japan at the cheap end of the global scale, which seems surprising given the relative wealth of the Japanese market (it is one of a handful of countries where Netflix has set a low price point).[13] Given the ultracompetitive nature of the Japanese streaming market, this was a carefully considered choice for Netflix.

Japan was already awash with digital media services, and had been for years prior to Netflix's launch. Given long commuting times in Tokyo and other major cities, Japanese consumers have long had a deep familiarity with mobile television and interactive services. Telcos and entertainment providers in Japan have developed clever ways of getting content to consumers, including subscription, ad-supported, and transactional models, often combined in one platform. Popular services included NicoNico, a popular YouTube-like site; Gyao!, an AVOD service part owned by Yahoo; Hikari TV, J.Com On Demand, and dTV (all distribution platforms linked to major telcos and cable providers); and streaming-service offshoots of media retailers (DMM.com, Rakuten Showtime). In the SVOD space, there was Avex, Amazon Prime Video, and Hulu Japan. The latter is a former subsidiary of the U.S.-based streaming service Hulu that was purchased by Nippon TV in 2014 and now runs independently from its American namesake, offering HBO content exclusively and a wide range of *anime* and Korean, Taiwanese, and Japanese dramas.

Netflix did not fill a clear market gap in Japan, where consumers are strongly inclined toward their own national shows. As Rob Cain (2015) noted in *Forbes*, this posed a problem for Netflix in Japan because "the billions of dollars that Netflix has invested in Hollywood movies and TV series, and in original programs like *Orange is the New Black*, *Unbreakable Kimmy Schmidt* and *Wet Hot American Summer* will have negligible value there." Instead, Cain predicted that Netflix "will have to spend (and probably overspend) for programming rights to Japanese language content that it will be unable to amortize anywhere else in the world." This proved correct. When Netflix launched in Japan, the company boasted that it had 40% Japanese content in its catalog (Shaw 2015), which was a lot higher than in many other Netflix territories—but it still was not enough for Japanese audiences. The local content deficit was compounded by the fact that Amazon Prime video had a much larger collection of Japanese content in its library, making Netflix seem uncompetitive and out of touch.[14]

This led some commentators to use the term *kurofune* (black boats) when reporting on Netflix's launch in Japan (as in *kurofune no nettofurikkusu*—Netflix of the black boats).[15] *Kurofune* is a historical reference to the gunboat diplomacy of Commodore Matthew Perry, who ended two centuries of Japanese isolationism by forcibly "opening up" Japan's markets to foreign trade in the 1850s. Used in this context, the term evoked the perceived threat of mass-scale importation of low-quality American content into Japan. While not a mainstream view in the Japanese media, the use of the *kurofune* discourse—including by some right-wing commentators—gives a sense of the political context surrounding Netflix and other foreign media services entering the Japanese market.

Netflix was especially criticized for its measly collection of *anime*, which is a must-have for any streaming service in Japan. To be fair to Netflix, this *anime* deficit was not entirely the company's fault. Netflix found it very difficult to license high-quality *anime*, reflecting its wider problem of securing rights to popular Japanese content. The major Japanese TV networks already had most of the rights locked up—Nippon TV had teamed up with Hulu, Hikari TV had a deal with Asahi Media Group, DMM partnered with Toyko Broadcasting Service (TBS), and so on. Netflix, with few options, eventually inked a deal with the remaining commercial broadcaster, Fuji Television, and now serves as the exclusive online distributor of Fuji-produced content. But securing licensing rights to popular content remains an issue for Netflix in virtually all countries, and this became a key driver of Netflix's decision to start producing its own content rather than licensing from uncooperative suppliers.

Netflix began coproducing with Fuji in 2015, releasing the reality show *Terrace House* and the coming-of-age drama *Atelier* (set in a lingerie company). *Hibana*, a show about aspiring stand-up comics, soon followed. In 2017, Netflix upped the ante by announcing it would release 20 new *anime* series of its own, including a reboot of the well-known 1980s series *Saint Seiya: Knights of the Zodiac*. Even bigger news followed when Netflix announced it was developing a new animated family series based on the ultrafamous Rilakkuma (literally *relax-bear*) cartoon character, akin to Hello Kitty in its stature inside and outside Japan. These recent developments mark a significant shift in Netflix's Japan strategy. In addition to claiming that these new investments would help them compete in the market, Netflix executives have also been talking up Netflix's ability to help these quintessentially Japanese shows reach an international audience. "Just as there are fans of Hollywood in Japan, there are die-hard anime fans in France, Brazil, the U.S. and all over the world," stated Greg Peters, president of Netflix Japan (Peters in Jarnes 2016).[16] He added: "There's this tremendous potential in Japan; so many stories—the manga, the novels—and now we have the opportunity to unlock this potential in a way that hasn't been able to be done before."

It is too soon to tell whether these new production commitments will represent a turning point for Netflix in Japan. Based on current trends at the time of writing, it seems reasonable to think that Netflix will remain popular with Anglophiles, expats, and hardcore TV buffs but may struggle to win over the masses given the fierce competition from local incumbents. In this uniquely inward-facing media market, Netflix may have to resign itself to remaining a niche service rather than a mass-market proposition—a solid out-come compared to the company's troubles in China.

China

China's scale cannot be ignored by any serious media company, and Netflix has long had its eye on the "world's biggest audience" (Curtin 2007) as a site for future expansion. Given the nature of Chinese media and its regulation, Netflix executives approached the market with caution and care. Rather than switching-on without permission as they did in most other countries, Netflix undertook a long engagement with Chinese regulatory agencies and local media companies, seeking a partnership arrangement that would let it operate legally in the PRC.

The early signs were not promising. By early 2016, many commentators were wondering whether Netflix would ever crack the Chinese market. In the global switch-on at CES, China was the only major country left off the new Netflix coverage map. (The other non-Netflix territories—Iran, North Korea, and Crimea—were affected by U.S. trade sanctions.) Hastings knowingly quipped during his CES launch speech that "the Netflix service has gone live in nearly every country of the world but China—where we also hope to be in the future," prompting laughter from the audience. But Netflix's China service was not to be. In October 2016, Netflix included the following passage in its third-quarter investor release:

> The regulatory environment for foreign digital content services in China has become challenging. We now plan to license content to existing online service providers in China rather than operate our own service in China in

the near term. We expect revenue from this licensing will
be modest. We still have a long term desire to serve the
Chinese people directly, and hope to launch our service
in China eventually. (Netflix 2016, 4)

These sentences spelled the end of its plan to launch a
Netflix China service. Soon after, Netflix announced a
partnership with the streaming site iQiyi (owned by the
search giant Baidu), which would become the official
home of Netflix original content in China. Netflix had
clearly decided there was no possibility of running its own
platform for the Chinese market. It was not the only
American company to come to this conclusion. Earlier in
2016, Uber had sold its China operation to rival ride-
hailing company Didi Chuxing.

To understand why Netflix failed to enter China, we
need to know something of the country's famously complex
and opaque internet regulations. State control over digital
platforms falls under the jurisdiction of at least three regulatory bodies, the Cyberspace Administration of China, the
Ministry of Culture, and the Cyberspace Affairs Council
of China, each of which issues its own decisions. Approval
from all three regulators is needed for a service to operate
safely in China. Further regulatory risk was added by the
fact that the official rules made no semantic distinction between subscription streaming services and live streaming
services (i.e., amateur webcam channels)—the latter being
a particularly sensitive area for the Chinese government,
which was cracking down on "anti-social" online activities.
Professional video streaming services risked being tarred
with the same brush as the live streamers.

Even if Netflix had been able to enter the Chinese market, it would have found it difficult to attract more than a

niche audience. As in India, cost was an issue: the standard Netflix global price of $7.99 USD/EUR was around four times what Chinese services such as iQiyi and Leshi were charging. Without lowering its price point, Netflix would have been out of reach of most consumers. Another challenge was the highly competitive nature of China's streaming ecology. Powerful incumbents—including video streaming platforms run by Baidu, Alibaba, and Tencent, the all-powerful "BAT" triumvirate that dominates China's tech economy—left little room for Netflix to make an impression. These platforms also have a distinctive style that Netflix could not easily compete with. Unlike Western portals, Chinese video services are multipurpose platforms that integrate free ad-supported video alongside many other services, including dating, shopping, real estate, and transport. Netflix's pure-play SVOD model would have been quite unfamiliar to most Chinese (and possibly somewhat boring). Netflix would also have inevitably faced the same local-content problem it encountered in other markets.

In any case, Netflix never got a chance to try its hand in China. The regulatory hurdles that blocked its entry remind us that global media is a domain of national policy as well as of commercial strategy. In this sense, the China case stands as a reminder that "the state remains the primary site of governance with the capacity to make decisions, assign resources and enforce laws within territorially defined societies, and they do so in the context of a global order that—in spite of multiple forms of supranational institutional and organizational innovation—remains primarily defined around a system of states" (Flew, Iosifidis, and Steemers 2016, 7).

Reading across these three case studies—India, Japan, and China—we can appreciate how difficult and messy the internationalization process is in practice for multiterritory digital services. Seen from this perspective, dynamics of demand in global television do not appear to have been fundamentally changed by digital distribution, in the sense that localization remains vital to explaining the success or failure of transnational television services. Netflix, despite its many powers, cannot easily overcome the locality of taste—it must *localize* itself if it wishes to compete on a global scale. Furthermore, it is by now apparent that global markets do not simply lust after Netflix in the same way that MTV thought the world might "want their MTV"—as a pure kind of obsession for a superior product. Instead, Netflix is in the more difficult position of needing to try and find its own niche within resilient and highly local taste formations wherever it goes.

In this sense, the story of Netflix is not entirely new; indeed, it closely resembles the history of transnational satellite channels expanding into Europe, Latin America, and Asia in the 1990s. The basic dynamics of taste and demand—what television audiences in different countries want from their television services—do not appear to have been fundamentally altered by digital distribution, confirming the home truths established by Chalaby and Straubhaar. Nonetheless, there are structural differences between an over-the-top television service like Netflix and the earlier broadcast and satellite TV companies. While Netflix faces the same market challenges as other transnational television companies, it has distinct ways of addressing them. Its long-distance localization model represents a Silicon Valley engineering response to a much older, and thoroughly cultural, business challenge—the stubborn locality of taste.

Content, Catalogs, and Cultural Imperialism

Que le carnage commence! (Let the carnage begin!)
—*Le Monde* newspaper, September 13, 2014, the day before Netflix's launch in France

In previous chapters, we have explored the technical infrastructure behind Netflix's international service and the commercial issues the company has faced in entering diverse markets. Now it is time to turn our attention to matters of content and to the vast array of movies and television series that the platform makes available to audiences around the world.

Netflix's internationalization has presented some fascinating content-related questions, many of which recall long-standing debates in media and communications research about the origin and intensity of international television flows. The most contentious issue across the board is the relative lack of local content within the platform compared to the abundance of U.S. programming. Netflix chief content officer Ted Sarandos has stated on several occasions that the company follows a "secret formula" when curating its international catalogs, involving "around 15% to 20% local [content] . . . with the [other]

Figure 5.1. Netflix: a global pipeline for U.S. content? Screenshot by the author.

80%, 85% being either Hollywood or other international content" (Netflix 2014; see also Block 2012). This definition of local appears to be quite flexible: in English-language markets, local might mean a mix of British and American content, whereas Spanish or Mexican dramas might count as local in South American nations, and Egyptian and Turkish soaps might count as local in the Arab world. Relatively few Netflix catalogs—especially those of small, minor-language markets—actually offer 20% local content from the country in question. Many smaller markets make do with a mishmash of whatever bargain-basement material has been licensed to Netflix under global terms. In any case, the bulk of the catalog in each country—including both licensed titles and original productions—skews heavily American.

Netflix makes no apologies for this. On its investor relations website (Netflix 2017c), the company explains that "local content represent[s] a minority of viewing in our markets." It explains its strategic and selective investment in original local content as "a way to onboard members and to introduce them to our global [i.e., U.S.] catalog," adding that "our aim is not to replicate the programming of the local broadcaster or TV network in a given market but to complement our service with local content where appropriate."

For countries with strong traditions of local content regulation, this is a potential problem. Netflix now competes in some markets with broadcast and pay-TV companies that are often obliged to screen a substantial amount of local, national, or regional content as part of the conditions of their license. Netflix, as an over-the-top service, is generally immune from such requirements. As a result, regulators, screen industry bodies, and civil society groups are increasingly looking closely at the Netflix catalog in their country and even commissioning their own studies of its content mix.

In Canada, where Netflix has been operating since 2010, research conducted for the Canadian Radio-Television and Communications Commission estimated that only 3.3% of feature films and 13.7% of TV content in the Netflix catalog were of Canadian origin (Miller and Rudinski 2012). In Australia, where Netflix launched in 2015, there were only 34 Australian movies in the catalog and around the same number of TV shows available upon launch (Scarlata 2015). More recently, research conducted by the European Audiovisual Observatory (Fontaine and Grece 2016) noted that across Netflix's 28 European catalogs, Hollywood movies typically account for over two-thirds of the film titles on offer. In the majority of European countries—including Austria, Bulgaria, Cyprus, the Czech Republic, Estonia, Finland, Greece, Croatia, Hungary, Lithuania, Latvia, Malta, Romania, Slovenia, and the Slovak Republic—there were no local films available on Netflix at all.[1]

As these figures suggest, the rise of Netflix presents challenges for existing media policies—especially local content policies designed in a broadcast era—that seek to maintain "some kind of dynamic equilibrium over time between locally produced media content and material sourced from overseas" (Flew 2007, 121).[2] Policymakers trying to keep up-to-date with the furious pace of technological change are increasingly looking to Netflix and asking questions such as: Is it a problem that Netflix is so Hollywood-centric? How does it compare on this front to local broadcast and pay-TV channels? Should Netflix be expected to screen and support local media productions?

Does the service provide a diverse "window onto the world," or is it yet another vehicle for U.S. domination?

Many of these questions involve normative assumptions about the relationship between national and global media and as such demand careful assessment. They also require us to think about the relative power of Netflix in particular national markets, and the sometimes divergent desires of audiences and policymakers in those markets. Approaching the Netflix catalog as an object of policy and politics, this chapter surveys some of the content-related controversies that have emerged in various countries and explores how media regulatory debates around this issue have been evolving in recent years.

Revisiting the One-Way Flow

The current discussion about Netflix catalogs is reminiscent of an earlier debate in international television research that may be familiar to many readers. Since the 1970s, media scholars, policymakers, and activists have been passionately arguing about the idea of cultural imperialism, a term referring to the wholesale export of media, information, and culture from the West to "the rest," and the relationship of economic and cultural dependence thus created. The idea first came to public attention in the late 1960s and 1970s through the work of Herbert Schiller (1969) and Armand Mattelart (1979), and in books like Allan Wells's *Picture-Tube Imperialism* (1972) and Jeremy Tunstall's *The Media Are American* (1977). It also gained institutional traction via UNESCO research and policy debates in the 1970s, as newly postcolonial nations across Africa and Asia asserted their cultural independence from former imperial powers.

For television studies, a legacy of the cultural imperialism debates has been the idea of the one-way flow. This is a much-contested notion, but the term itself simply refers to the argument that the United States (and to a lesser extent the United Kingdom and Western Europe) dominates the global television trade through the mass export of movies and TV series into foreign markets, including the deliberate dumping of content at cut-rate prices. As the theory goes, U.S. exports provide a cheap way to fill local schedules while also instituting a relationship of dependence between center and periphery. The end result is that *CSI* and *Grey's Anatomy* can be seen almost anywhere, while flagship productions from the rest of the world travel less frequently beyond their national borders.

The one-way flow has been a divisive topic within television studies, and while the theory has been questioned on conceptual and empirical grounds, it persists as a powerful meme in global policy discussions. For this reason, when thinking about the issue, it is necessary to distinguish between explanatory theories and actually existing discourses that shape policy and practice. The key is not so much whether the one-way flow idea was right or wrong but rather the fact that many people continue to *feel* there is a one-way flow, which requires an understanding of where this idea came from and how it circulates in the context of contemporary discussions about internet-distributed television.

A key document in this regard is the 1974 UNESCO report *Television Traffic—A One-Way Street?* by Kaarle Nordenstreng and Tapio Varis. This famous study examined broadcast TV schedules in 50 countries to establish an evidence base about the origin and direction of cross-border flows—culminating in the authors' well-known diagnosis

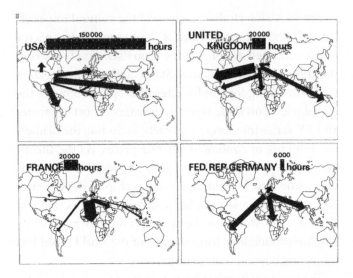

Figure 5.2. The distribution patterns of the major program exporting countries in the early 1970s, showing annual exports in hours. The width of the arrows is proportionate to the share of the total export flow. Source: Nordenstreng and Varis (1974, 30).

of a one-way, West-to-rest pattern.[3] The report used a simple infographic to illustrate this finding (Figure 5.1). These conclusions were confirmed by additional UNESCO studies that appeared over the next decade (Larsen 1990; Mowlana 1985).

While raising tricky definitional issues when it came to determining the national origin of coproductions or news programs with inserted material, this approach provided an empirical basis for strong claims as to the direction and intensity of international television flows. Similar conclusions were reached in the major 1977 study *Broadcasting in the Third World* by Elihu Katz and George Wedell, who concluded that "the flow of entertainment programs and entertainment formats is almost wholly in one direction," adding that "the flow of news is even more unidirectional" (Katz and Wedell 1977, 166). While their assessment of the cultural impact of television exports was carefully qualified, Katz and Wedell were clear in their argument that developing nations (in the 1970s at least) were structurally reliant on cheap imports from the United States and on a "homogenized brand of popular culture either copied or borrowed from broadcasting in the West" (vii).

The basic method for international television studies in the 1970s was analysis of broadcast schedules. But such research also embodied a critical argument about the global order as it seemed to be emerging at the time. Taken to its logical extreme, the one-way flow could be seen as one element in an imperial project that extended longer histories of political and economic domination. Television functioned as a microcosm of a larger truth: the continued domination of the developing world by means economic, political, and cultural. This way of thinking, which Miller et al. (2005) would later describe as the "Global Effects Model," was clearly of its time—a product of Cold War politics and the political turbulence associated with decolonization, postcolonial nationhood, and the emergence of the nonaligned bloc in the United Nations and UNESCO. As Jean Chalaby observed, it was also a product of Marxist dialectics in that it "transposes the Marxist interpretation of social classes in perpetual conflict to the relation between place and culture" (Chalaby 2009, 231).

Television export studies were fundamentally about the *distribution*, not reception, of content. Like today's research on Netflix that shows the measly amounts of local content in national catalogs, the 1970s studies could measure the presence of American content in global schedules but could not account for the relevant prominence, popularity, or cultural impact of such content. Nor could they explain what American TV meant to people in their everyday lives. This led many scholars to question cultural imperialism's hypodermic-needle theory of reception and its romantic notion of the nation as a besieged space of cultural purity (Tomlinson 1991; Miller 1992). Barker (1997, 182) went so far as to claim that cultural imperialism had become "an increasingly inadequate concept for understanding television under contemporary conditions." Meanwhile, Straubhaar (1991) and Sinclair, Jacka, and Cunningham (1995) offered revisionist accounts foregrounding the scale of interregional television trade, especially in Latin America. Their research established that many nations replaced imported content with local production as their media systems developed, which suggested that American content was less monolithically dominant than originally thought. Under the weight of this new evidence, U.S. imports came to be seen by some scholars as "a kind of televisual Polyfilla, plugging the gap in the schedule" (Tracey 1985, 22), rather than an unstoppable force of cultural domination.

In rehearsing these arguments, I make the obvious point that export power does not translate directly into cultural power. At the same time, at least some aspects of the cultural imperialism thesis remain important for understanding how global audiences feel about digital media services like Netflix. The dominance of U.S. media is still an empirical fact that must be reckoned with, as Netflix programming shows. Concerns about the one-way flow have never gone away, and inequalities in television trade are a live and legitimate issue—regardless of how we account for their cultural effects. Audiences still want to watch TV in their own languages (and complain when they cannot do so), and policies still exist in many countries to protect local content and nurture local stories. All of this suggests that the basic conditions for public concern about one-way flow remain broadly intact.

Milly Buonanno is a scholar who has carefully negotiated this apparent contradiction between the explanatory deficiencies of the one-way flow and the persistent anxieties of national media policy. Buonanno's research on European television programming in the 1990s and the following decade revealed that 50 % of the movies and drama series shown on terrestrial TV in Europe were North American. Nonetheless, she is extremely careful about making claims on the basis of such evidence. Consider this passage from her marvelous book *The Age of Television*, which provides a nuanced account of export dynamics that carefully qualifies the idea of U.S. dominance:

> It may indeed be true on the supply side that "the media are American" (Tunstall 1977), but this assertion is subject to the distinctions mentioned . . . : not television in general, but television drama; not as a permanent state of affairs, but subject to variations in time and space. But on the consumption side things are different. . . . We do not need to espouse the theory that the public has become saturated with too much American television, although this may be true. We should rather uphold the structural nature of the gap between supply and demand for the foreign product, which has not come about by chance. (Buonanno 2007, 94)

Following this logic, the challenge of explaining international television flows is not so much about picking one paradigm over another (globalization vs. cultural imperialism) but rather about making careful distinctions between distribution and reception, economic structure and audience/buyer agency, and the more specific dynamics of various program types. Buonanno's other key insight here is that while the paradigm of cultural imperialism "has lost its central position in the field of explicit theories," it "still maintains its hold at the level of the implicit theories that underlie public discourse and ordinary conversation" (87). Many small nations still feel the specter of cultural imperialism even if they use a different vocabulary to express it, and whether or not the perceived threat comes from the United States, France, India, Japan, China, or Korea.

How does this apply to Netflix? It is impossible to understand the international policy landscape around internet-distributed services, especially SVOD, unless we heed the lessons of the cultural imperialism debate. As SVOD services command an increasing share of global television viewing in many nations, concerns about the one-way flow on unregulated and mostly U.S.-based platforms are increasing. This involves a renaissance of cultural protectionism in many nations and a reassertion of the regulatory power of the nation-state (Flew, Iosifidis, and Steemers 2016). The advent of internet television has somewhat unexpectedly served to *intensify* long-simmering arguments about television flows in general, precisely because of internet-distributed television's increased capacity for "programming from afar."

Netflix Catalogs and Media Policy in Europe

The European Union's undiminished commitment to the "cultural exception"—the idea that "creations of the spirit are not just commodities" and "the elements of culture are not pure business," in the words of former French president François Mitterand—has shaped continental media policy at a deep level. A central goal of the European Union's overarching media law, the Audiovisual Media Services Directive (AVMSD, 2010), is to nurture a "European audiovisual space." The AVMSD, which replaced the former Television without Frontiers Directive (1989), thus seeks to ensure that content from EU member states flows liberally across national borders within the region while preserving access to local-language and national-language film and television content in specific countries: Dutch films for the Dutch, Spanish television for the Spaniards. To this end, current EU law provides minimum standards of European content (at least 50% of broadcast television content must be European), which individual member states may increase if they wish.

The AVMSD—which is under revision at the time of writing—has until recently been vague about on-demand services, referring only to a general obligation to "promote" European works.[4] European member states have been left to make their own rules about how this might happen (for example, through content quotas, levies, or investment requirements). Quotas now apply to video-on-demand services in a number of European nations, though the specified percentage of content varies considerably. For example, VOD services based in France, Austria, and Lithuania must dedicate at least half of their catalogs to European content, while services based in the Czech Republic, Malta, and Slovenia face a much lower hurdle (10%).[5] Broadly speaking, Central and Eastern European nations have tended to favor lower or no quotas, while Western European nations opt for more stringent quotas.

In addition to these rules about the presence of European content in on-demand catalogs, some countries have also imposed specific rules about how digital platforms present their content to viewers (European Commission 2014).[6] In Romania, it is required that providers indicate the country of origin of titles in the catalog. In Poland, rules specify how platforms should promote European content in trailers, home pages, and catalogs. Estonian law requires that providers give some prominence to recent European works, those released within the last five years. These laws add another layer of complexity to the European content quotas that already exist in many EU nations.

The most regulatory-minded country in the EU on cultural policy issues is France. Here, interlocking policies exist to protect national customs, language, and media from Anglophone influence. France, which has chosen to ramp up the minimum EU content quotas to a higher level, presently requires that 40% of broadcast television and radio programming must be of French origin, and no less than 60% of on-demand content must be European. In addition, free-to-air TV, pay-TV, and TVOD services are obliged to invest a portion of their revenue in European and French original productions, and strict chronology rules regulate a film's movement through nontheatrical windows (Blaney 2013; O'Brien 2014; Keslassy 2013). France also has specific rules about discoverability in on-demand services. Current law specifies that "the home / front page of the service provider must display a substantial proportion of European or French-speaking works" and that visuals and trailers of these works must be shown, not just the title (European Commission 2014). These rules apply only to services based *in* France and have no effect on services based in other countries, even when they cater to French audiences. Under the AVMSD's "Country of Origin" principle, providers must satisfy only the rules of the country in which the service is registered. Hence, Netflix's European operations, which are headquartered in the Netherlands, only need to satisfy the lax Dutch requirements about European content, no matter the size of the audience in France. As a result, Netflix and other providers have an incentive to set up their operations in less regulatory-minded member states.

European regulators are well aware of this issue, which results in uneven compliance with the spirit of European content rules across the continent. In response, they have been seeking to harmonize national rules to create a more uniform regulatory environment. To this end, the European Commission announced in 2014 that, as part of its Digital Single Market strategy, it would begin consultations for a new version of the AVMSD. Three months of intense lobbying followed, during which major European telcos and U.S. tech companies (including Netflix and Verizon) argued the case for retaining the status quo, while European screen industry groups (and a number of EU member states) argued for stronger regulation. France was especially vocal in its submission to the EC: it wanted a new content quota to promote cultural diversity and to protect European audiovisual industries from unfair competition from American companies. In its submission, France also attacked the tendency of "certain non-European companies" to set up shop in EU member states with light-touch regulation:

> This strategic relocation, associated in particular with digital services, is a form of "forum shopping" and ultimately takes place at the expense of the value chain of European

content. Thus publishers established in France—who are subject to ambitious regulations to support the audiovisual industry, protect creators and offer diversified and quality content to viewers—are in direct and unfair competition with the services established in another country, which are not subject to the same rules but generate profits in the same market. (Government of France 2015)[7]

Unsurprisingly, Netflix came out strongly against these proposals. Its submission to the European Commission warned that "rigid numerical quotas risk suffocating the market for on-demand audiovisual media services" and that "an obligation to carry content to meet a numerical quota may cause new players to struggle to achieve a sustainable business model." Netflix even offered some free policy advice to Brussels, suggesting that "the focus of European audiovisual media policy should be on incentivising the production of European content and not imposing quotas on . . . providers who may then struggle to meet the supply" (Netflix 2015c, 16–17).

After much deliberation, the European Commission eventually came down on the side of France. Its proposed new AVMSD package, released in March 2016 as "a media law for the 21st century," contained some very significant new regulatory requirements that are likely to reshape the audiovisual landscape in Europe (at the time of writing, these have yet to be implemented).[8] The most controversial requirement was a 20% European content quota in VOD and SVOD catalogs (subsequently increased to 30% by the European Parliament), designed to create "a more level playing field in the promotion of European works." Importantly, these rules would apply regardless of the country of origin: major European *and* foreign services like Netflix would all have to comply if they wished to do business on the continent. A second element of the new AVMSD applied to the recommendation systems used by digital video platforms. Platforms would be required to give European content a reasonable degree of visibility within their catalogs, even if this meant rewriting algorithms to prioritize such content. These discoverability requirements are significant, signaling the EC's intention to ensure providers could not work around the content quota by licensing cheap, low-quality European content and burying it in their catalogs.

EU regulators were merely acknowledging the obvious fact that users of Netflix and similar services do not experience the catalog *as* a catalog or as a static list or schedule but rather as a series of interactive, personalized recommendations that are algorithmically sorted according to user viewing history, demographic, and location data. Given the importance of recommendations in driving viewing behavior, these systems are clearly important objects for cultural policy regulations to consider. If implemented, this second rule would require changes to the way Netflix's algorithms work, given that content recommendation is tailored to an individual's viewing history.

Naturally, Netflix pushed back on the filtering proposals, noting that they are "not compatible" with how people use SVOD services, and claiming they would undermine Netflix's ability to recommend what people are most interested in. Netflix had previously argued to the EC that personalized recommendations were the best kind of promotion for EU works, precisely because they are based on demonstrated preferences and will therefore recommend more EU content to people already known to like such content. In Netflix's words, "the consumer personalization approach allows for more European works to be available to subscribers that are interested in such types of content and helps them find those European works—in effect, promoting them and creating a virtuous demand-supply cycle" (Netflix 2015c, 17).

Netflix's talk of a "virtuous demand-supply cycle" may be gilding the lily, but they do have a point about recommendation. In SVOD services, the catalog is really the raw inventory of content from which selections are algorithmically drawn and is rarely visible to users in its entirety. The algorithm has considerable agency here. However, this also means that regulators' arguments can be considered valid when they insist that the algorithm is a legitimate target for regulation. This tension reflects the fuzzy boundaries of today's media policy debates, which are not just about *how* and *how much* to regulate but also *what* to regulate (i.e., which objects, technologies, and practices count as "media"). The EU, for its part, seems to be moving in the direction of treating algorithmic recommendation as a media system in its own right, or at least an appropriate surface for regulation.

At the time of writing, these new regulations have yet to be legislated. Nonetheless, the fact that these proposals are being debated at the highest level is testament to the level of concern about these issues within Europe, where questions of national sovereignty, cultural diversity, and, indeed, cultural imperialism are still taken quite seriously. In bringing discoverability into the equation, the AVMSD foretells a new approach to media regulation—an attempt to retrofit existing regulatory models for the more complex, context-dependent landscapes of digital distribution—and suggests some of the techniques that future governments may use to intervene in the digital media economy.

The Canadian Situation

The EU's interventionist position in these debates is worth noting as an exception to the general deregulatory thrust of global media policy in recent decades. Relatively few governments have the resources or appetite to enforce these kinds of measures. Others are discouraged from imposing national content quotas on digital platforms because of preexisting commitments in trade agreements—Australia, for example, traded away this right during negotiations for the 2004 U.S.-Australia Free Trade Agreement.

Canada—a midsize country with a long history of cultural policy protections for its bilingual media space—is an interesting case. With nationally funded public-service media institutions, Canada has long-established systems of state support for cinema and television production and a quota system for Canadian content (Cancon), including requirements that VOD services include Canadian movies in their catalogs.[9] Netflix has a relatively long history in the country, having operated there since 2010, and 53% of English-speaking Canadian households are now thought to subscribe (Robertson 2017). By this measure, Netflix would have to be considered a mainstream rather than a niche media service in Canada. Yet, as a digital service, it is not required to invest in Canadian production in the same way as television and pay-TV companies are. Progressives in Canada argue for tighter regulation of Netflix as a way of preserving Canadian culture in the face of unfair over-the-top competition. A report by the Canadian Centre for Policy Alternatives gives a sense of this position:

> Canada is no stranger to turning points in the media landscape, and Ottawa has proven its ability to respond in ways that continue to support Canadian cultural production. In the 1930s, for example, we created the CBC to counteract the domination of the English-language airwaves by U.S. radio. In the 1960s, when American television programming flooded Canadian airwaves, we wrote new Canadian content regulations. Today the issue is once again domination, but this time of the Internet by mostly U.S. OTTs, which have become their own distribution networks as well as their own creators of new media. (Anderson 2016, 9–10)

Here the content regulation question once again becomes entangled with questions of national sovereignty. The argument is that unless Canada moves urgently to regulate internet television services, the country will be "in danger of losing control of its broadcasting system and of causing severe damage to the production and delivery of Canadian cultural programming"—something the report describes as "akin to the unregulated radio era of the 1930s and 1940s, when Canadian airwaves were dominated by American programming and we were merely a branch plant location for U.S. culture" (44). This vision of Netflix as yet another U.S. media company muscling into the national media space is familiar to Canadians. As a journalist from the *Globe and Mail* wrote recently, "Canadians' digital habits are stripping millions out of the system as the dollars spent on ad buys, Netflix subscriptions and Amazon purchases

TABLE 5.1. Paid subscriptions to over-the-top services in
Canada (English-speaking population estimates)

Netflix	53%
CraveTV	9%
Amazon Prime Video	5%
Sportsnet Now	5%

Data source: Media Technology Monitor Survey, cited in Robertson (2017).

all flutter down to California. Meanwhile, Canadian content risks drowning in the Internet ocean" (Taylor 2016).

On the other side of the political spectrum, Canadian conservatives are skeptical of the quota system and
of media regulation in general (Globerman 2014). They
argue that imposing new taxes or obligations on online
services is unnecessary for quality Canadian content to
thrive, and that such regulation stifles digital innovation
and free markets (seen as the best vehicle for creating culture). Conservatives, like progressives, have also found
reasons to bring Netflix into the political debate. During
the 2015 Canadian federal election campaign, prime minister Stephen Harper released a scripted video on Twitter
promising that he—unlike Justin Trudeau—would never
tax over-the-top services. "I love movies and TV shows,"
he tweeted, "I'm 100% against a #Netflix tax. Always have
been, always will be #NoNetflixTax."

From 2013 to 2014, the Canadian Radio-Television and
Telecommunications Commission ran a public consultation process called *Let's Talk TV* to guide the future of
national television policy. Hearings were held in 2014, and
Netflix representatives were called to give evidence. However, Netflix executives would not divulge commercially
sensitive data requested by the commission—such as the
number of Canadian subscribers and Netflix's dollar-figure

investment in Canadian content—other than to claim that "Canadian content is thriving on Netflix." In response, the commission struck Netflix's evidence from the public record as a rebuke for the company's recalcitrance. This fuss, which made a few headlines in Canada at the time, gives an insight into Netflix's sometimes insouciant attitude toward national media regulators. Ultimately, Netflix did not need to play by the CRTC's rules and was able to opt out of the process when it was expedient to do so (Pedwell 2014; Anderson 2016, 36; Wagman 2017).[10]

The relationship between Netflix and Canadian regulators has grown more complicated over time. In 2017, while announcing the recommendations of the government's new Cancon review, heritage minister Mélanie Joly made two major announcements concerning Netflix. The first piece of news was that Netflix and the Canadian government had signed off on a deal for Netflix to establish a production base in Canada and to invest C$500 million in Canadian screen production over five years. Netflix claimed this was new money as opposed to already-committed funds; however, some observers were skeptical of the claim, and others asked whether the funds would be used to produce runaway U.S. productions, taking advantage of the lower production costs over the border. Netflix (2017b) insisted that they "invested in Canada because Canadians make great global stories"—a response that did not seem to sway its critics (Taylor 2017). The second piece of news was that the government had decided not to proceed with plans to impose a sales tax on digital streaming services. Despite announcing these two commitments on the same day, both Netflix and the Canadian government stressed there was no link between them and that the production spending was not given in return for the tax break.

Nonetheless, the timing of the announcement raised many eyebrows.

Comparing policy debates in the EU and Canada reveals some differences as well as some similarities. Canada, unlike the EU, is not proposing a Netflix quota, though the issue of Cancon on Netflix remains highly controversial. Canada's current audiovisual policy emphasizes "strengthen[ing] the creation, discovery and export of Canadian content," suggesting a more strategic subsidy model for telling Canadian stories, in contrast to the European quota model. The term "discovery" connotes distribution, search, and algorithmic filtering—all the characteristics of the Netflix era. Canada, like the EU, is putting digital distribution front and center in its thinking and is attempting to retrofit long-standing media policies around this theme. Whether this will be enough to satisfy Canadian content creators and cultural policy advocates remains to be seen.

Do Audiences Actually Want Local Content (on Netflix)?

We have heard a lot from policymakers and policy advocates, but what about audiences? How do they feel about the local content issue? The voice of the audience does not feature prominently in local content debates and, as a result, it can be difficult to know how local they want their television to be and how this sentiment might vary across different kinds of television services.

In thinking about this issue, it is worth going back again to the *Television Traffic—A One-Way Flow?* report and to some comments in the report's appendix from Elihu Katz. Writing more than 40 years ago, Katz identified some important questions that are still highly relevant to the Netflix case. He began by noting that critiques of U.S. or European dominance in broadcasting rest on "certain value assumptions which deserve to be made explicit":

> There is the assumption, first of all, that national creativity—that is, home-made programmes—is a good thing. Some people may disagree: it is possible to argue that nation-states ought not to be striving to create their own cultures or to continue their own parochial traditions. Perhaps it is better to be more cosmopolitan. It is certainly cheaper to let Sam do it.
>
> A related question is what kind of free-flow of information do we want? Underlying this study is the assumption that an egalitarian flow in all directions is an ideal toward which to strive. But perhaps some people would be satisfied with today's more unidirectional flow, if only because they prefer not to tamper with information. Perhaps the global village is not so bad? Perhaps the nation-state will wither away more quickly if television and other cultural media are homogenized! Perhaps conflict among nations will be reduced if they find common ground in *Peyton Place*. Perhaps the audiences everywhere do really prefer *Bonanza* to locally made cultural products? (Katz, in Nordenstreng and Varis 1974, 47)

Swap *Peyton Place* and *Bonanza* for *Stranger Things* and you see how Katz's point applies to Netflix. The assumption that audiences necessarily *want* to see their own stories on Netflix more than they want to see Hollywood stories is questionable because the wider national media landscape may already be structured and regulated to provide significant amounts of local content through existing broadcast and pay-TV channels. The issues Katz identifies here clearly remain unresolved.

It is difficult to speak in general terms here. Every country has its own story when it comes to local content. In Chapter 4, we considered the situation in India, where Netflix is very much a niche service. India is not having the same kind of policy debate about "Indian content on Netflix." With a very mature and successful commercial screen industry of its own—where language functions as a natural protection measure—India has no real need for Netflix to add to the existing stock of national self-representation. Instead, Netflix's signature brand of edgy English-language content appeals to those audiences already predisposed to imported content—and everybody is more or less happy with that arrangement, at least until Netflix becomes popular enough that it starts to unsettle existing incumbents. But the story is different in Anglophone countries such as Australia, New Zealand, and the United Kingdom. In these countries, American programming can have a stronger substitution effect on local programming, because of the language factor. Many Australians, Kiwis, Brits, and Canadians are more than happy to trade up their local dramas for higher-budget English-language Netflix programming.

Here we can see how the cultural policy debates vary significantly from country to country. Inherent differences in national markets shape the potential impact of Netflix within those markets. An additional complicating factor is that Netflix occupies different *market niches* within those countries. In the United States, Netflix is a mainstream middle-class product, affordable and accessible to most people with a high-speed internet connection. In other parts of the world, Netflix use is either negligible or limited to cosmopolitan upper classes whose tastes may not be indicative of those of their fellow citizens. (Recall that the company explicitly targets "English language-speaking elites" in foreign markets [Netflix 2016].) In other words, Netflix caters to different *kinds* of global users, as well as different national markets, so how regulators feel about Netflix (the perceived threat level) varies accordingly.

This raises some fascinating regulatory questions. If Netflix is effectively a niche service, serving only a small number of viewers, should it still be asked to represent the local way of life in its programming? Perhaps this is an unfair expectation to begin with. After all, Canadians do not expect European art house movie SVOD services like Mubi to support local production, because they have very small audiences. Nor would a New Zealander expect a Chinese video-on-demand service to support Kiwi productions just because that service is potentially available in New Zealand. This is why media policies designed for convergent media landscapes often make a distinction between media services on the basis of size and audience rather than delivery mechanism. For example, the EU's Audiovisual Media Services Directive is clear about the fact that it is designed for "mass media" and that "low turnover companies, thematic services and small and micro enterprises are exempted from these requirements." Netflix certainly does not have a low turnover when considered as a global operation, but its turnover is low in specific countries. Is this enough to qualify Netflix for regulatory immunity? Should regulatory definitions of size be based on the receiving-country presence or the overall size of the corporation? These are some of the policy debates that the case of Netflix brings into focus.

In this context, calls for Netflix to behave more like a national television service may be somewhat naïve, and possibly counterproductive. As Katz observed many years ago, we must accept the possibility that many global viewers prefer *House of Cards* to their local shows (or rather, they like the option to watch it *as well as* their local shows). This is what Netflix chief content officer Ted Sarandos thinks, at any rate. Speaking on a quarterly result call to financial analysts in 2014, Sarandos made the following observations about how his 80/20 rule was working in Western Europe, where Netflix had recently launched:

> One of our first indicators that we are getting the mix right is how many hours of viewing people are participating in. And in France and Germany, the viewing hours are quite healthy relative to all of our other launches. So . . . the consumers are finding the things they want.
>
> The tricky thing is figuring out, is the local content something that people want in the long term. Because when we first go at a new market, I think people are mostly excited about those things that they didn't have access to before.
>
> So *Orange Is the New Black* was by far the most watched show in both France and Germany, and in fact all of the markets that we launched. So it tells you that with all the differences in taste, . . . they all rallied around that show. (Netflix 2014)

It is in Netflix's interest to say this, and we should not place too much stock in a company's public relations. But nor should we dismiss Sarandos's argument out of hand. Implicit in his comments are certain claims: that Netflix subscribers mostly want international content as opposed to local content (though I doubt Sarandos would say this of North American Netflix subscribers) and that premium content can be transcultural in its appeal. With access to the Netflix black box, Sarandos may be in a position to know, but it is worth emphasizing how controversial such claims still are in the context of media globalization debates.

Is there a way to reconcile these positions? On one level, the problem remains intractable, because actually existing audience activity rarely aligns neatly with cultural theory. But if we look once more to the history of debate around global television flows, we see a path through the fog. The trick is to use *both/and* rather than *either/or* thinking. Audiences do not choose between the local and the global but combine both in their everyday lives; they move between these scales of identification, at different times and for different purposes. Audiences understand that local programming is good for some things (news, sports, comedy, reality TV), while American imports are often better for other things (high-end drama, spectacle, thrills). And, of course, the sheer predominance of a particular kind of content in the SVOD catalog does not determine viewer experience as much as it would in a linear schedule, precisely because the outer reaches of the catalog can be conjured to appeal to and appear to those users who have already expressed an interest in watching more obscure content, whether Scandi-crime, Hindi blockbusters, or Nollywood (Nigerian) movies (all of which can presently be found in Netflix catalogs).

Netflix's specificity for international television research therefore lies not in the issues it raises about cross-border content flows—because we have seen this before—but in the fact that it can effortlessly combine the local and the global *within the one platform* and constitute itself as many different products simultaneously because of the magic of algorithmic filtering. Hence, while its political economy seems to be an extreme case of the local/global dialectics that much media globalization theory is based on (North American enterprise coming over the top of the national), its internal constitution as a platform blurs these boundaries in new ways. This may not be enough to redeem the platform in the eyes of regulators, who are likely to remain focused on the number of national titles in the catalog, but it should act as a prompt for television scholars to develop new ways to think about the articulation between catalogs, recommendation algorithms, and national media policies as we move further into an on-demand media environment.

6

The Proxy Wars

When considering how internet-distributed television has evolved globally, it is important to take into account the many informal user practices that have developed alongside, and in interaction with, the major platforms. Let me begin by offering a personal story that explains why this issue is significant for understanding Netflix. Like many TV fans in Australia—where Netflix was geoblocked until late 2015—I first experienced Netflix not as a local service but as a U.S.-based service that had to be accessed covertly by using a virtual private network (VPN). During these years of nonavailability between 2010 and 2015, several hundred thousand Australians covertly signed up for the U.S. Netflix service using a credit card, a fake U.S. residential address, and a VPN or other proxy service.[1] As long as our VPNs were active when we signed in, we could experience Netflix in the same way as Americans do. This workaround provided many happy hours of streaming until Netflix introduced an antiproxy policy in early 2016.

Australia was not an isolated case. In the early years of Netflix's internationalization, use of VPNs and proxy services was common in many countries—including Mexico, Canada, New Zealand, France, and Britain—where a local

Figure 6.1. Marketing for Getflix, one of the many DNS proxy and VPN services that facilitated unauthorized cross-border streaming (May 2016). Screenshot by Chris Baumann.

Netflix service was not available or where the local catalog was perceived as inferior to the U.S. version. Countless YouTube tutorials and websites offered step-by-step
instructions on getting around geoblocking, making this a
fairly mainstream practice. While all this was against Netflix's terms of service, the company did not seem to mind
having the extra paying customers, and it all seemed like
harmless fun.

My point here is that the history of Netflix as a global
platform cannot be understood only as a tale of Silicon
Valley innovation and international market entry. It is
also, inevitably, a history of user experimentation, circumvention, and copyright infringement. These unauthorized
practices are not just margin notes around the edges of
the Netflix story; they are integral elements underlying
the growth of Netflix as a global media service. Equally, the
policies developed by Netflix to curtail this activity—the
"proxy wars"—also form an important part of the wider
institutional history of internet-distributed television.

User Practices and Platform Policies

In their influential work on Twitter, new media scholars
Jean Burgess and Nancy Baym (2016) develop a "platform biography" approach to understand how platforms
change over time. This involves attending not only to
the features and design of a platform but also to how
the platform is experienced, adapted, and transformed
by its users. In the case of Twitter, it is well known that
many popular user features—such as adding a hashtag
to posts—were invented by users rather than platform
designers. For Burgess and Baym, this raises the question
of how everyday digital practices can "emerge . . . through

user experimentation, as people seek to concretize the platform's emerging uses and norms, and in some cases to develop tools to enhance and better coordinate these conventions" (Burgess and Baym 2016, 10).

Extending this way of thinking to Netflix, we can start to appreciate the delicate back-and-forth between platform design and user activity that is a feature of most digital media. How have people variously used, adapted, and in some cases tricked the Netflix service? What tools and technologies have they employed to do this? Netflix is a relatively more closed platform than Twitter, but it is nonetheless amenable to a range of unofficial user practices. These run the gamut from innocuous platform hacks to more serious transgressions.

At the minor end of the spectrum, we find activities like *password sharing*, where users share their login credentials with friends, family, or strangers. This is a common practice: 40% of Netflix subscribers in the United States have reportedly let other people use their logins (Wallenstein 2013).[2] Other examples include uploading custom subtitles to Netflix or installing Chrome and Firefox browser extensions that add extra features to the Netflix website, such as IMDB (Internet Movie Database) ratings, random-play functions, microgenre browsing, or enhanced personalization. These user practices are uncontroversial and widely tolerated.

In contrast, geoblocking circumvention has proven to be a more troubling issue for Netflix and for entertainment industries generally. To understand why this is so, we need to know a little about the technology and business of geoblocking. This begins with the humble IP (internet protocol) address, the set of numbers assigned to a device that is used to send and receive data online. IP address "lookups" are a simple, cheap, and very widely used way to geolocate customers. Various free and proprietary databases have been developed for this purpose. Leading providers such as Akamai and Maxmind offer automated country-level and city-level geolocation databases, costing a fraction of a cent per query.[3]

From a commercial perspective, digital media platforms use IP geolocation because it offers a cheap and easy mechanism for market segmentation, personalization, and legal compliance (Svantesson 2004; Goldsmith and Wu 2006; Trimble 2012, 2016). Streaming services will typically check a user's IP address to confirm the user is in an authorized service zone. Outside this zone, the user will be confronted with an error message or an endlessly buffering screen.

IP geolocation is an imperfect system with many limitations, the most important being that geolocation can only tell you about the IP address of the device rather than the physical location of the person using it. Despite modest improvements over time, the system remains open to manipulation. As an Akamai representative has stated, IP geolocation "isn't meant [for] people are who trying to be evasive. . . . It's meant for the 99 percent of the general public who are just at home surfing" (Associated Press 2004).[4]

For the remaining 1%, various technical solutions exist to circumvent geoblocks and gain out-of-region access to online services (Lobato and Meese 2016). The most commonly used tools are VPNs, Smart DNS (domain name system) proxies, and free browser add-ons. VPNs, which can be used for privacy and business purposes as well as for circumvention, typically cost around US$5–$15 per month and provide an encrypted tunnel to a remote

TABLE 6.1. Popular DNS services used by Netflix's international subscribers, and their marketing slogans (circa 2015)

Service	Marketing Slogan
Unblock.US (DNS)	"Unblock Everything on Netflix, Spotify, Hulu and More"
uFlix (DNS)	"Expand your Netflix library!"
Proxy DNS	"Netflix, Sling, HBO, Hulu, and more . . . Outside USA."
Unotelly (DNS)	"Freedom. Security. Flexibility."
Blockless (DNS)	"Your Internet. Your Freedom."
MediaHint (DNS)	"Content Unblocked—Countries have borders. The Internet shouldn't."
Getflix (DNS)	"Unblock Netflix and Hulu Plus FREE with our 14 day trial"
Torguard DNS	"Unblock content anywhere"
TV Unblock	"American DNS codes"
Unlocator	"Watch Netflix anywhere"

server. There are hundreds of VPN suppliers in the marketplace, including well-known brands such as Private Internet Access, Hotspot Shield, and HideMyAss. Smart DNS proxies are cheaper than VPNs, costing a few dollars per month. They will effectively mask your IP address but do not encrypt your traffic. Finally, free browser add-ons such as Hola and MediaHint are easily installed and much simpler to use than VPNs or proxies. Users select from a list of countries, then choose an available video streaming service (e.g., selecting U.K. in Hola then allows the user to select BBC iPlayer).[5]

For simplicity, the rest of the chapter will use VPN as an umbrella term for these various circumvention tools, even though they are all technologically distinct. The next step in our analysis is to understand how Netflix responded to the rising popularity of these tools. We will then consider how the company's policies changed over time, and identify the strategies and values that motivated these changes.

Historicizing Netflix's Shifting Policies on Geoblocking

Netflix's internal policies on VPN use can be divided into roughly three periods. The first phase of internationalization, between 2010 and 2014, was characterized by a relatively permissive attitude. The second phase, between 2014 and 2016, witnessed growing external pressure to adopt a stricter policy. Finally, in 2016, Netflix introduced a new VPN-detection technology and recommitted to geoblocking as a principle.

During the first phase, Netflix was only available in the Americas and parts of Europe. Users outside these regions would see the message "Sorry, Netflix is not available in your country yet," and many used a VPN to get around this block. It is impossible to tell how many users accessed Netflix using VPNs at this time, but the practice was sufficiently well known for *Variety* to refer to Netflix's "black market diaspora" (Wallenstein 2014). The limited research literature also gives clues as to the cultural drivers of VPN use in various countries. Vanessa Mendes Moreira de Sa (2016) notes that tech-savvy Brazilian Netflix subscribers used VPNs to access the U.S. Netflix service because it offered English-language closed captions not available in Brazil (which was important for students). Studies by Leaver (2008), Beirne (2015), Stewart (2016), Shacklock (2016), and Lobato and Meese (2016) also show the prominence of geoblocking and circumvention in various other countries.

Netflix preferred not to comment publicly on VPN use at this time. The company was busy building a global brand with global market awareness. In fact, cultivating tech-savvy early adopters was part of its long-term strategy. Netflix enjoyed a reputation as a company that understood the internet and its users. It was reluctant to turn away paying customers, who helped to inflate the company's U.S. subscriber numbers and share price.

The second phase of Netflix's VPN policy was characterized by intense industry pressure. Rights-holders were starting to get anxious about what they saw as wholesale parallel importation or, worse, piracy. Tense conversations took place between Netflix, its suppliers, and its competitors. Concerns about VPN use were publicly aired in the trade papers and tech press, and were amplified by the publication of a number of reports (some rather speculative in nature) about the scale of the VPN "problem." One report by Global Web Index, "The Missing Billion," estimated that 28% of its global sample had used VPNs—amounting to "419 million people in GWI's 32 markets" (Global Web Index 2014, 9).

Armed with these statistics, many rights-holders pressured Netflix to take a tougher line on VPNs. "I know the discussions are being had . . . by the distributors in the United States with Netflix about Australians using VPNs to access content that they're not licensed to access in Australia," stated Simon Bush, CEO of the Australian Home Entertainment Distributors Association (Bush in Reilly 2014), "They're requesting for it to be blocked now, not just when it comes to Australia." By late 2014, all eyes looked to Netflix for a solution to the perceived VPN problem. The collateral damage from their laissez-faire approach was starting to mount as grumpy rights-holders continued to air their grievances.

The tension ratcheted up a notch in November 2014, when WikiLeaks released an archive of emails from Sony Pictures—including a leaked memo from Sony Pictures' chief digital strategy officer Mitch Singer, dating from December 2013—that revealed the depth of feeling within the studio about Netflix's lack of action on VPNs.[6] Noting that "this is a politically and emotionally charged issu[e] with Netflix," the Sony memo concludes that "Netflix can and should do a much better job geofiltering." Other leaked emails criticized Netflix's geolocation as "very leaky" and lamented its "reluctance to address this issue."[7]

Rights-holders like Sony saw circumvention as a problem for at least two reasons. The idea of consumers wantonly "stealing" content from out-of-region services was naturally upsetting because it undermined the ideal of an orderly digital marketplace. VPN use seemed like an affront to the whole intellectual property system, and especially to the idea of territorial market segmentation—a foundational concept of copyright. Industry lobby groups began to speak of "VPN piracy," equating circumvention (which is actually more akin to parallel importation than piracy) with the specter of illegal downloading.

The second, more tangible concern was that VPN users were starting to dilute the value of content rights. If Canadian or British Netflix users could use VPNs to watch a particular program via the U.S. Netflix catalog rather than paying to watch it on a local Canadian or British pay-TV service, then major rights-holders (especially the Hollywood studios) would not be able to demand the same prices for their content that they were used to charging, because they could not guarantee territorial exclusivity. Whichever way you looked at it, rights-holders and distributors both had a lot at stake in territorial market segmentation. The geoblocking circumvention problem was showing just how crucial market segmentation was to digital media business models.

Asked about VPNs during a Netflix quarterly results call in April 2015, Ted Sarandos tried to play down the disquiet among suppliers. "Yes, [VPN use is] one of the many things that we have discussions with studios about on an ongoing basis, and we do continue to work with them, and work with the VPNs," Sarandos stated. "To be honest with you, it's kind of a whackamole to get ahead of the different usage of VPNs. It's become kind of a lifestyle thing for a very small segment of the population" (Sarandos in Netflix 2015a, 8). Rights-holders were not mollified by these remarks, and the pressure continued to build throughout 2015.

Finally, on January 14, 2016, Reed Hastings announced the global switch-on at CES in Las Vegas. Shortly afterward, Netflix issued a press release—"Evolving Proxy Detection as a Global Service"—announcing the introduction of an industry-standard geoblocking policy. Noting that Netflix's new status as a global service had removed the need for out-of-region access workarounds, the press release re-stated Netflix's commitment "to respect and enforce content licensing by geographic location," adding that "we look forward to offering all of our content everywhere"—a reference to the company's goal of achieving global licensing terms with its suppliers.

Shortly after the press release went out, internet forums lit up with commentary, criticism, and skepticism. Was Netflix serious about blocking VPNs? What technology would they use to do so? What would happen to the VPN industry, which thrived on consumer demand for cross-

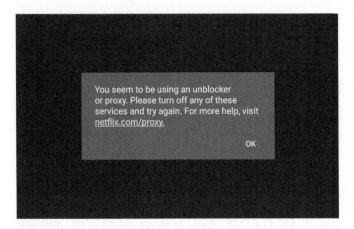

You seem to be using an unblocker or proxy. Please turn off any of these services and try again. For more help, visit netflix.com/proxy.

OK

Figure 6.2. Netflix error message for VPN users. Screenshot by the author.

border streaming? For the first few weeks, it appeared that nothing had changed. Most VPNs were still working as usual. On internet forums and social media, many gloated that Netflix's new VPN blocking system had failed. However, by the end of February 2016, most VPN users were seeing an unfamiliar and unwelcome message, shown in Figure 6.2.

Over the next few weeks, countless reports appeared on social media about VPN users being blocked from the U.S. catalog. Virtually all major VPNs seemed to be affected. Netflix's new proxy detection system was building up a dynamic database of IP addresses it determined were associated with VPNs and then blocking them all. Of course, there were still various ways to fool the system (VPN providers could actively change their IP address ranges, and tech-savvy users could set up a premium VPN subscription with an individually assigned IP address).

However, it was all starting to look too much like hard work to many users. Some decided it was not worth the hassle.

Meanwhile, VPN companies scrambled to come to terms with the new policy. Private Internet Access and Mullvad took the opportunity to distance themselves from circumvention, noting that they never endorsed such activity. Other VPN companies stuck to their guns. Torguard insisted it could still outsmart Netflix with its premium dedicated-IP service. The company's CEO stated that: "We greatly expanded our Dedicated VPN IP pool and now offer Dedicated IP options in over 55 countries worldwide. This has proven to work flawlessly for users who wish to bypass VPN blockades with geo-restricted streaming services" (Ernesto 2017a). Other providers quietly intimated that they would be able to work around Netflix's new policy. An Express VPN representative rather ambiguously stated "the first rule of Netflix is: do not talk about Netflix" (ibid.).

DNS proxy services were hit especially hard by Netflix's policy shift, because they relied heavily on the hardcore streamer user base. Several companies disappeared or morphed into VPN providers. Others experienced technical challenges. My colleague Chris Baumann and I tested a range of these DNS services in 2016 and discovered that, while the majority of them could still allow access to the U.S. Netflix catalog, they were not always reliable. Few services were consistently effective as circumvention tools, meaning that circumvention was now a fairly time-consuming activity that was likely to appeal only to the most committed users.

While certainly not bulletproof, Netflix's anti-VPN technology has been more or less effective in its stated aim. The blocking of VPNs changed the public perception of circumvention, making it appear difficult and bothersome instead of quick and easy. It also mollified the suppliers on whose content Netflix was absolutely reliant. In this way, Netflix was able to contain the perceived threat and redefine VPN use as a niche activity for hardcore geeks rather than mainstream internet users.

Making Sense of the Policy Shifts

Aside from the whack-a-mole games, what does the history of Netflix's VPN policy tell us about the relationship between user practices, technological restrictions, and company policies?

The digital media business is inherently leaky because it is built on the sale and leasing of access to infinitely reproducible goods, such as digital videos and ebooks. History tells us that what people do with these digital goods cannot easily be controlled, no matter how strong the digital rights management, so the imperative for forward-thinking media companies is not necessarily to stop all informal use of their property but rather to extract as much value as possible from a leaky market.

Netflix understood this well. Until 2016, its response to the VPN problem was not punitive. It was not about shutting down informal uses of its system. Rather, it was about extracting the maximum value from VPN users. This is why Netflix dragged its feet and carefully timed the introduction of its anti-VPN policy to align with the global switch-on—even though it had already known about the circumvention problem for years and rights-holder concerns had been growing for some time.

The trigger for the policy shift was economics rather than ideology. At a certain point, it made commercial sense for Netflix to stop thinking like a new-economy Silicon Valley company (committed to enhancing user experience through innovation) and to start thinking like an old-fashioned media company (by aggressively protecting its rights). The logic of the market dictated a cultural change in the company's values and self-identity. Netflix transformed from a "friend of the geeks" into an intellectual property defender because it made commercial sense to do so.

As further evidence of this shift, consider how Netflix's policies on illegal downloading have evolved in recent years. In public statements up until 2015, Reed Hastings took a moderate position on piracy, avoiding extreme antipiracy positions in favor of a pragmatic attitude that emphasized the importance of converting piracy into paid consumption. In 2013, Hastings stated:

> Certainly there's some torrenting that goes on, and that's true around the world, but some of that just creates the demand. Netflix is so much easier than torrenting. . . . We don't even think about trying to get rid of it. What we really think about is how to build an awesome service that people just want to use. (Hastings cited in Schellevis 2013)

Public statements like this—of which there are many on record by Hastings and other Netflix executives—suggest market realism rather than copyright puritanism. At this point in its history, Netflix wanted to present itself as a company that was reasonable, forward-thinking, and understanding of the internet and its users.

This relatively permissive attitude changed once Netflix became a major-league content producer. In recent years, Netflix has been aggressively enforcing its copyrights by

Reed Hastings on piracy and VPNs

April 2015: "The key thing about piracy is that some fraction of it is because [users] couldn't get the content. That part we can fix. Some part of piracy however is because they just don't want to pay. That's a harder part."

April 2015: "[VPN-enabled viewing is] certainly less bad than piracy. It's not something we encourage. It's actually very hard to detect, because VPN gets very good at covering their tracks for all the obvious reasons. And because we're focused on getting global very quickly, I think we'll see this issue disappear, and it will disappear because we'll be able to meet the demand directly in all the countries."

June 2015: "Well, you can call it a problem, but the truth is that [piracy] has also created a public that is now used to viewing content on the Internet. . . . We can think of this as the bottled water business. Tap water can be drunk and is free, but there is still a public that demands bottled water."

October 2016: "We've been very successful at finding technological ways of inhibiting the cross-border VPNs, which is roughly, like I'd mentioned, we didn't win the bidding for the Disney movies in the UK, so it's clearly not fair to allow our UK subscribers to watch the Disney movies from Canada or to the US. And so we found, with the help of the studios, some more technology that enforced their rights."

Sources: various press reports; Netflix quarterly earnings call transcripts 2015Q1 / 2016 Q3

sending out more than one million takedown notices to pirate websites (Google 2017). At the same time, it has expanded its legal team to include more copyright attorneys with antipiracy expertise. A March 2017 job advertisement for a Global Copyright Protection Counsel position at Netflix's Los Angeles office gives a sense of this work. In charge of "industry-wide anti-piracy strategic initiatives and tactical take down efforts with the goal of reducing online piracy to a socially unacceptable fringe activity," the counsel would be responsible for providing "detailed landscape and piracy trends analysis"; spotting "new trends and changes in the ecosystem"; lobbying; and providing "outreach" to pirate sites, sharing platforms, and social media services (Ernesto 2017b). In January 2018, Netflix and its partners in the Alliance for Creativity and Entertainment—including Amazon, HBO, BBC, and the Hollywood studios—also started filing lawsuits against suppliers of pirate streaming boxes (Ernesto 2018).

Antipiracy "education" was also part of the agenda. In 2017, Netflix also released a memorable antipiracy promo on YouTube, targeted at the French market. The subtitled video featured four *Narcos* cast members from the Cali Cartel, who threaten viewers with all manner of unpleasant deeds should they access *Narcos* illegally. "Hey you," intones Pêpê Rapazote, who plays the menacing character Chepe in the series, "Do you think we didn't see you Googling '*Narcos* season 3 download'?" Other cast members offer various warnings, such as, "If you want your entertainment, if you want your show, you gotta pay the Cali Cartel, *hijo de puta*," and "There is no please, no por favour, no *s'il vous plait*. . . . There's bullets for you, your family, and all the people you send to watch *Narcos* on those shitty websites full of pop-ups *sucios* [dirty]."

This was a new, humorous take on the old antipiracy advertising formula. While it stands in sharp contrast to Netflix's previous statements on piracy, this position makes sense when we consider what Netflix had at stake in its original content investment. By 2016, Netflix was spending billions of dollars on original content production each year. The company was now a major rights-holder, and it was starting to act like one—by introducing

Hollywood-style content protection and antipiracy policies. This investment in original content now colors every
aspect of the company's strategy. Netflix wants to recoup
this multibillion-dollar cost, enforce its rights, and minimize leakage in the system. Having orderly territorial markets and an effective antipiracy strategy is essential in this
regard.

Cultural Consequences of the Proxy Wars

In this chapter, we have seen how Netflix progressively seg-
mented its international markets into defined territories
while users variously accepted, resisted, or circumvented
this segmentation. A few years out from these events,
and with the benefit of hindsight, we are in a position to
answer some longer-range questions about these turbu-
lent years: How did the "proxy wars" shape the evolution
of television streaming as a global media practice? What
did hundreds of thousands of internet users learn from
the experience of using VPNs to access Netflix? Do these
geoblocking battles have any wider relevance to internet
culture?

It seems to me that one of the key legacies from these
years has been an increased public awareness of the geography of digital markets. During the geoblocking and VPN
debates, people started asking questions that showed some
of the amusing inconsistencies of the copyright system.
("Why is this series available in Albania but not Alberta?"
"Why can't I watch my favorite show online even when I'm
happy to pay for it?") Many internet users experiencing
geoblocking on a regular basis also came to form views on
related issues like international price discrimination and
windowing. In some countries, these concerns translated

directly into government policy, as policymakers sought to constrain "unjustified geoblocking."[8]

Any user of a VPN or similar service during these years would have become familiar with marketing slogans promising "borderless TV" or the ability to "watch TV like a local." Meanwhile, Netflix catalog comparison sites such as AllFlix, FlixList, and Flixsearch encouraged users to consume media beyond their national borders. These websites were brazen about drawing attention to the disparities in the catalog system and promoting their own services as a workaround. A new popular discourse had emerged, characterized by a logic of cross-border comparison of digital media services and contempt for the principle of territoriality.

Catalog differences have eroded over time as more Netflix originals have been produced, but they remain significant in many users' minds. Unlike music streaming platforms, which enjoy global licensing terms, and social media sites, which are full of user-uploaded content that is not typically georestricted, Netflix was a global service with an obviously territorial catalog system. As such, it became a stalking horse for all the failings of territorial copyright generally, even though it arguably did more than most other companies to minimize them. In other words, Netflix came to stand in for a wider set of problems that were not of its making.

It is easy to dismiss all this controversy about VPNs and geoblocking as a first-world problem, and in some senses this is true: access to new-release movies and TV series is a privilege, not a human right. But to do so would also be to miss the subtle consequences of the events outlined in this chapter. The desire for a borderless Netflix inevitably helped to acquaint early adopters with digital rights and internet privacy discourses promoted by VPNs. It fostered a popular awareness of what are otherwise obscure technical matters. The common experience of geoblocking, leading to a desire for circumvention, operated like a "gateway drug" for a wider set of political issues.

The love affair with VPNs and cross-border streaming could be perceived as a degraded form of popular cosmopolitanism, in the sense that it involves a desire to cross borders and come into contact with media systems (or servers at least) in far-off lands. At the same time, this cosmopolitan impulse was also a symptom of cultural imperialism, because mostly what people were looking for when using VPNs to access Netflix was new-release American content. Regardless, it is safe to say that the rise of global Netflix helped foster in users a vernacular awareness of the geography of copyright and the contradictions of digital markets. The proxy wars may be over for now, but this genie cannot easily be put back in its bottle.

Conclusions

When Reed Hastings stood on stage at CES 2016 and announced that Netflix had been switched on in 130 new countries, he evoked an image of effortless market entry—the annihilation of space by digital technology. The reality of digital markets as shown in this book is rather more complex. Netflix, like other multiterritory television operators before it, has learned that global reach is not the same thing as global popularity. Digital distribution does not come "over the top" of culture; it must negotiate the rough terrain of markets characterized by fundamental differences in tastes, values, cultural norms, viewing habits, income levels, and connectivity.

The regulatory blowback that Netflix has confronted as it has established itself in different countries has been more intense than anticipated. Looking forward, there is a strong possibility of more aggressive regulation and quotas on the cultural policy front, especially in the EU and the Anglosphere. We can also expect that concern about Netflix's unlicensed status will become contentious in countries with strong traditions of state media control. Meanwhile, the wider political backlash against Silicon Valley may lead to further scrutiny of Netflix's tax and accounting practices.

Taking all this into account, it seems that the idea of digital markets as borderless, "flat" spaces of exchange and consumption is misguided. As Netflix discovered, the reality of doing business in hundreds of markets simultaneously is messy and difficult, and governments are actively pursuing ways to extend national sovereignty online. Internet delivery has not resolved the political tensions inherent in running a global media business; indeed, it seems to be *intensifying* existing political debates.

Old and New Lessons

There are several lessons to be learned (or relearned) here for television studies. As I have argued throughout this book, the story of Netflix's internationalization and eventual localization is unique in many ways, but it also confirms a few home truths about global television that have already been established in previous research by Liebes and Katz (1990), Chalaby (2005, 2009), Buonanno (2007), Straubhaar (2007), Pertierra and Turner (2013), and others. These lessons can be summed up as follows.

AUDIENCES STILL SKEW LOCAL IN THEIR TASTES Netflix's localization strategy and its commitment to new original production in multiple languages underscore the fundamentally local nature of global taste. While there are pockets of demand for high-touch English-language content in many nations, the general pattern is that audiences still want television in their own language, with familiar faces and culturally relevant stories. The urbane English-language dramas and comedies that Netflix has specialized in so far are not enough to build a mass-market television service on a global scale. Demand for local-language news, talk shows, reality television, and other local genres will continue to be satisfied primarily by national television providers who have a long history of producing this kind of content and a deep understanding of what their audiences want. Netflix cannot—by definition—be the future of television in all countries.

THE GLOBAL DOES NOT DISPLACE THE LOCAL, BUT THE TWO CAN HAPPILY COEXIST While Netflix cannot replace mainstream television at global scale, it can certainly exist as a successful niche service. As various satellite channels have shown, catering to a small user base that finds a lot of value in imported media can still be a profitable endeavor if extrapolated across many countries. The conceptual implication here is that the relationship between local/national and global/U.S. content is often complementary rather than substitutive: the latter is a desirable extra layer of content that appeals strongly to certain kinds of viewers. Netflix's offering can therefore be highly valued *alongside* local-language talk shows, news, and other staples, because they satisfy different and complementary desires.

THERE IS NO COHERENT NETFLIX EFFECT AT GLOBAL SCALE Rather than having a uniformly disruptive effect, Netflix has had quite different effects in different national contexts—ranging from disruption of broadcast and pay-TV incumbents (as in Canada, Australia, and other English-language markets), to modest success as a niche service (in much of Europe and Latin America), or no impact at all (in Africa and the Middle East, for example). Netflix's subscriber base now differs substantially from country to country precisely because Netflix occupies different market positions within those countries.

UNIFORM GLOBAL PRICING "PRICES OUT" LOW-INCOME CONSUMERS Netflix's decision to maintain a broadly consistent price point from country to country, calibrated to suit the incomes of westerners, means it must remain a niche service in most low-income countries. Recall the situation in India, where Netflix's user base is English-speaking urban elites. Compare this to North America, Australasia, and Western Europe, where Netflix is a cheap, mainstream offering enjoyed by a more diverse group of users. Its low price point in these markets compared to average wages means it is more accessible to low-income groups and is a better value than buying or renting DVDs. In other words, Netflix is a different service and a different value proposition in each market. The conceptual implication here, building on the previous point, is that we need to be careful about speaking about a typical Netflix audience, a typical Netflix user, or a coherent Netflix effect. These generalizations are useful for analysis but obscure a more important insight, which is that Netflix has had *differentiated effects* on the markets it has entered. This brings us to a final point with an important conceptual implication.

NETFLIX HAS BEEN STRUCTURALLY TRANS-FORMED BY ITS INTERNATIONALIZATION Netflix is not the same service worldwide: catalogs, language options, and platform features change when accessed from different countries. Looking forward, the general trend is toward more differentiation rather than less, because of increasing regulatory pressure for Netflix to behave differently in different countries. Taking this into account, in some senses it may now be more appropriate to see Netflix as a collection of *national media services* tied together in the one platform rather than as a uniform global service. Consequently, the recent history of Netflix is not just a story of a singular service encountering the frictions and differences of the global market; it is also the case that Netflix is becoming internally differentiated as it moves into and through these new markets. "How is the object transformed—and how does it transform—from stage to stage, context to context?," Scott Lash and Celia Lury (2007) ask in their study of global brands. The Netflix case provides some answers to this question, suggesting that the accrual of difference is an inevitable consequence of a brand's movement through the world.

Streaming beyond Netflix

In the previous chapters, I have endeavored to use Netflix as a thinking tool—a platform, if you will—to explore the evolving relationship between television and global media in a context of internet distribution. At the start of the book, I noted that Netflix is a specific service with unique lines of development. It does not stand in for internet-distributed television as a whole. Netflix is presently the world's largest SVOD platform, but SVOD is only one kind of service alongside AVOD, TVOD, hybrid platforms, and set-top box systems. In other words, Netflix is a small, but very powerful, part of a wider ecology. What happens when we provincialize Netflix and shift our focus to the rest of the ecology?

One way to answer this question is to look to China, where a very different streaming economy has emerged in recent years. In China, the streaming economy is dominated not by pure-play SVOD services but by multipurpose video platforms, such as Youku and Tencent Video, that integrate professional scripted content with news, shopping, education, games, investment advice, and other services. At the same time, the staggering rise of Chinese live streaming—amateur live channels hosted on video platforms—points toward an alternative future for internet-distributed television that lies somewhere between unscripted reality television and YouTube, with dashes of gamer culture thrown in.[1]

Live streaming is about everyday intimacy rather than high spectacle, distraction rather than immersion—a casual, low-intensity kind of television that stands in sharp contrast to Netflix's immersive, scripted-content offering. Within this emerging paradigm, the Netflix model appears rather quixotic. Even on its home turf, Netflix is quite unlike the other major platforms. YouTube, for example, is a multipurpose, mostly free service based on advertising. Amazon Prime Video is a loss leader for an e-commerce platform that also distributes individual linear channels for cord-cutters. Facebook is different still: a social network that is still experimenting with its video model. For this reason, the idea of a coherent "big tech" sector—exemplified in the "FAANG" stock category (Facebook Amazon Apple Netflix Google)—can be deceptive because each platform is imagining and shaping the future of television in distinct ways.

Seen in this context, Netflix's unwavering commitment to scripted narrative, spectacle, and old-fashioned televisual pleasure seems very old-fashioned (Wolff 2015). Netflix, at this stage of its evolution at least, is all about professional content; it has no interest in user-generated content. Its platform design encourages a cinematic mode of viewing. There are no opportunities for social interaction. As Amanda Lotz (2017a) argues, Netflix may well be structurally more similar to HBO and other subscriber-funded portals (and even to historical precursors such as lending libraries and subscription publishing) than it is to other digital video platforms. Hence our analysis needs to take into account the internal heterogeneity within the category of internet-distributed television. These parallel lines of development between SVOD, social media, and live streaming may eventually converge, but I suspect they are just as likely to keep diverging—with the effect that the ecology will become more complex and differentiated as it ages.

A final trend that I have observed while writing this book is the parallel growth of internet-distributed *linear* television services (live channels as opposed to on-demand libraries). These come in formal, semiformal, and informal variants. In particular, I have been fascinated by the rise of what is known as IPTV piracy, or low-cost internet-distributed subscription services offering hundreds of live satellite channels—a new and distinctive "global television" experience. Usually delivered through grey-market Android set-top boxes, pirate IPTV services offer a distinctive experience that is completely different from the way Netflix imagines television. Pirate IPTV is linear, live, and full of ads. Netflix is nonlinear, on-demand, and immersive. Both are global but in different ways: IPTV aggregates diverse international *channels*, while Netflix aggregates diverse pieces of content into a curated *database*. This is a very different vision of the future from what Netflix offers, reminding us of the heterogeneous nature of internet-distributed television.

There are other examples we could point to, although it would take another whole book to do them justice. My overall point here is that the evolutionary path of television from this point onward will not be unidirectional. There will continue to be many different models of internet-distributed television—each with their own geography—just as there have been many competing visions of television in the past. Within this evolving international context,

Netflix stands as a singular but fascinating case study—a specific vision of what global television might mean in an internet age—as well as a reminder that digital distribution cannot easily overcome the stubbornly local dynamics of culture, consumption, and taste.

ACKNOWLEDGMENTS

Like any television program, digitally distributed or otherwise, this book would not exist without the contributions of a great many people. Between 2015 and 2017, I was fortunate to receive an Australian Research Council Discovery Early Career Researcher Award (DE150100288), which allowed me to spend much of my time conducting research for this book and its companion volume *Geoblocking and Global Video Culture* (2016), which I coedited with James Meese. Over the course of this fellowship, I was lucky enough to work with a number of very talented scholars—Alexandra Heller-Nicholas, James Meese, Tessa Dwyer, Alexa Scarlata, Ben Morgan, Chris Baumann, Thomas Baudinette, Wilfred Wang, Ishita Tiwary, and Renee Wright—who provided research assistance as well as countless critical conversations. Briefing papers written by Thomas, Wilfred, Ishita, and Renee were especially important for Chapter 4. Naturally, I am also grateful to the Australian Research Council for making this research possible. Jonathan Gray, whose knowledge of television knows no boundaries, provided encouragement and advice over several years as this project took shape, and kindly hosted a visit to the University of Wisconsin-Madison in 2017, where I was able to try out some ideas that would make their way into this book. Amanda Lotz has been a generous interlocutor throughout and has helped me see this topic in entirely new ways. Colleagues in the Global Internet Television

Consortium provided a forum for comparative insight. Julian Thomas, Jennifer Holt, Hannah Withers, Patrick Vonderau, Stuart Cunningham, Joshua Braun, Scott Ewing (greatly missed), Jock Given, Tom O'Regan, Ellie Rennie, Rowan Wilken, César Albarrán-Torres, Dan Golding, Liam Burke, Jenny Kennedy, Esther Milne, and Aneta Podkalicka, among others, offered sage advice, support, and help with translations. I am also grateful to audiences at the Porting Media conference (Concordia University, October 2017), the Cultural Studies Association of Australia, and the International Communication Association conferences where various drafts of this work were presented. Finally, thank you to Eric Zinner and Dolma Ombadykow at New York University Press; Critical Cultural Communication series editors Aswin Punathambekar, Adrienne Shaw, and Jonathan Gray (again); and two reviewers who provided careful and constructive feedback.

This book draws on some ideas and material that were initially developed in the earlier publications *Geoblocking and Global Video Culture* (2016); "Rethinking International TV Flows Research in the Age of Netflix" (2017); "The Friction of Digital Markets" (2017); and "Streaming Services and the Changing Global Geography of Television," in *Handbook on Geographies of Technology*, edited by Barney Warf (2017).

NOTES

1 The video of Hastings's presentation can be viewed at www
.youtube.com/watch?v=l5R3E6jsICA.

2 This was the rule, but there were many exceptions—witness the
spillover of broadcast signals across national borders, especially
in Europe (discussed in more detail in Chapter 2).

3 YouTube fits in this category, too, but it is more properly
described now as a hybrid site offering a subscription package
(YouTube Premium), free videos, and paid transactional rentals.

4 The MAVISE database is available at mavise.obs.coe.int.

5 China Global Television Network (CGTN), formerly CCTV
International, is China's 24-hour English news channel.

6 These policies were documented in a famous slideshow (available at www.slideshare.net/reed2001/culture-1798664) that has
been viewed 17 million times and is now taught in management
schools.

7 I refer here to Wagman and Barra's Cultures of Netflix panel at
the 2016 ECREA (European Communications Research and
Education Association) conference.

CHAPTER 1. WHAT IS NETFLIX?

1 This historical progression can be studied via the company's
quarterly earnings releases since 2002, available at ir.netflix.com.

2 Think here of how Netflix wants to compete with international
pay-TV services while simultaneously insisting that it should not
be regulated like a media company—a topic we return to later.

CHAPTER 2. TRANSNATIONAL TELEVISION

1 An example is Time Warner CEO Steve Ross's 1990 speech
praising the electronic media's capacity for "free flow of ideas,
products and technologies in the spirit of fair competition" (cited
in Morley and Robins 1995, 11).

2 Note the trajectory implied in the subtitle, which gives a clue as
to Straubhaar's argument.

CHAPTER 3. THE INFRASTRUCTURES OF STREAMING

1 A note on sources: there is a considerable amount of technical material available online, as many of Netflix's engineers use open platforms such as YouTube, Slideshare, and Github to share internal information, as is common among tech circles. Useful sources include Netflix's Github repository (github.com/Netflix); the Netflix UI engineering, Netflix OSS, and Netflix Performance and Reliability Engineering channels on YouTube; and online videos from tech industry events where Netflix engineers regularly give keynote talks, such as AWS re:Invent. The Netflix Tech Blog is another rich resource, which gives a sense of how Netflix projects itself to the tech community at large.

2 There exist numerous lists of this kind (e.g., Jackson et al. 2007; Star and Ruhleder 1996) and many more studies that could be mentioned as exemplars. The list of terms and references here includes those I think are most useful for media and culture scholars.

3 Netflix CEO Reed Hastings is also well connected on the progressive side of U.S. politics and is a major donor to the Democratic Party.

4 In Africa, the most popular applications in terms of downstream internet traffic are YouTube (19%), general web browsing/HTTP (18%), Facebook (9%), and BitTorrent (8%) (Sandvine 2016b).

5 An AWS region includes various server locations within it, although it is difficult to pin down their geography with any accuracy. As a journalist from *The Atlantic* discovered when she tried to find out exactly where the AWS servers in Northern Virginia were located, "Unlike Google and Facebook, AWS doesn't aggressively brand or call attention to their data centers. They absolutely don't give tours, and their website offers only rough approximations of the locations of their data centers, which are divided into 'regions.' Within a region lies at minimum two 'availability zones' and within the availability zones there are a handful of data centers" (Burrington 2016).

6 Open Connect is for larger ISPs that have 100,000 users or more.

CHAPTER 4. MAKING GLOBAL MARKETS

1   Murdoch initially partnered with MTV, carrying MTV on his Star satellite service in Asia. He later broke off the agreement and developed his own clone service, Channel [V].

2   Individual broadcasters may use a combination of these strategies for different markets and combine country-specific channels with regional feeds suitable for a wider geolinguistic market. Bloomberg, when expanding into Europe, decided to develop country-specific channels for France, Italy, Spain, the United Kingdom, and the Germanic countries, whereas Fox Kids pursued a more expansive localization strategy (note the importance of dubbing for children, who may not read subtitles), with dedicated channels for France, Greece, Italy, Poland, Spain, the Netherlands, the German-speaking countries, Scandinavia, the United Kingdom and Ireland, Hungary and the Czech Republic, and Eastern Europe (Chalaby 2005).

3   In recent speeches, Hastings has repeatedly used the words "listen" and "learn" in reference to Netflix's international strategy, suggesting that the company is now moving along the localization knowledge curve described by Chalaby.

4   As an example, Taiwanese customers are billed, improbably, through Amsterdam.

5   Netflix executives have also acknowledged studying BitTorrent traffic as a proxy for demand in international markets (Schellevis 2013).

6   One example is the multinational firm Deluxe, which has offices in seven countries. Suppliers can be viewed at the Netflix Preferred Vendors website: npv.netflix.com.

7   The GILT industry has its own conferences (LocWorld, which hosts three international events per year), magazines (*Multilingual*), associations (Globalisation and Localisation Association, Localisation Industry Standards Association, etc.), and even a journal (*International Journal of Localisation*). Major companies in the GILT sector include Lionsbridge, CSOFT, Viacom, and VSI. The GILT professions are commonly referred to by the numeronyms L10N, I18N, and G11N.

8   One localization expert working at Netflix recalls the following anecdote in a localization podcast: "Last year . . . in Korea it was

pointed out, not very subtly, by a Korean journalist that the font that we [Netflix] were using in our artwork was not appropriate, and was not good enough. This was pointed out to our CEO in a press conference. It wasn't great for us in localization" (Jentreau 2017).

9 DFXP files can be created from the more common SRT-format files via various free websites. Free Chrome and Firefox add-ons offer users a similar solution.

10 The transcript of this conversation can be seen at np.reddit. com/r/india/comments/5lwg89/hey_rindia_netflix_employee_ here_just_wanted_to/.

11 Another asked, "When can we expect all sorts of payment options like all debit cards and others? For this exact reason, I'm unable to use Netflix."

12 This is the price for Amazon Prime membership, spread over a year.

13 The Premium Netflix package, however, is in line with the USD price.

14 Amazon Prime Video had 70% Japanese content at launch (Hadfield 2015).

15 An example can be found at matome.naver.jp/ odai/2144270998283023201 (courtesy of Thomas Baudinette).

16 Peters would later be promoted to chief product officer in April 2017.

CHAPTER 5. CONTENT, CATALOGS, AND CULTURAL IMPERIALISM

1 While the range of television (as opposed to movie) content on Netflix is somewhat more localized—around half the TV series in each catalog are American, and around a third are European (mostly British, French, and German)—TV series from smaller European countries are still largely absent from the Netflix platform (Fontaine and Grece 2016). New research is constantly appearing; for example, a 2017 study of the Australian Netflix catalog puts the level of local content at around 2% (Lobato and Scarlata 2017). Given the dynamic nature of catalogs, these studies need to be seen as part of an evolving research enterprise; there is still a lot we do not know about how catalogs change over time. Nonetheless, the general implication here regarding the lack of local content in most national Netflix services is fairly clear.

2 Flew is invoking Philip Schlesinger's notion of "communicative boundary maintenance" here.

3 Nordenstreng and Varis (1974) used a questionnaire to ask TV broadcasters around the world about their programming, and they supplemented this with their own analysis of publicly available TV schedules.

4 Article 13 of the AMSVD requires that on-demand services "promote . . . the production of and access to European works." One form of promotion specified in the directive is "the share and/or prominence of European works in the catalogue of programmes offered by the on-demand audiovisual media service."

5 The approach used in the French-speaking areas of Belgium is notable for its rather more complex approach, which makes reference to a very wide range of factors, including the presence of "specialized" and "thematic" collections within the catalog and the terms under which European content is licensed (how long films stay in the catalog).

6 A summary of current legislation can be found in the European Commission document "Promotion of European Works in Practice," available at ec.europa.eu/newsroom/dae/document .cfm?doc_id=6296.

7 My translation from the original French.

8 See ec.europa.eu/digital-single-market/en/revision-audiovisual-media-services-directive-avmsd.

9 The existing quotas for Canadian movies in VOD services, as specified in Broadcasting Regulatory Policy 2014–444, are as follows: at least 5% of English-language films, at least 8% of French-language films, and at least 20% of all other programming must be Canadian. The VOD category does not include over-the-top TV services like Netflix.

10 Google also refused to provide this commercially sensitive data to the CRTC.

CHAPTER 6. THE PROXY WARS

1 This figure, based on a private industry report commissioned in Australia, was very widely reported in press coverage throughout 2014 and 2015. Its accuracy cannot be verified, but neither was it

ever really questioned by industry stakeholders, most of whom seemed to think it was roughly on target.

2 One analyst, Michael Pachter from Wedbush Securities, estimated at the time that Netflix may have as many as ten million unauthorized users accessing its service via shared passwords (Wallenstein 2013). For his part, Reed Hastings (Netflix 2016 Q3 call) has stated that password sharing "is something you have to learn to live with, because there's so much legitimate password sharing like, you know, you sharing with your spouse, with your kids," adding that "there's no bright line" (i.e., clear distinction between legality and illegality) here. Hastings went on to note that password sharing was not a significant issue when it came to revenues and therefore not a major concern to the company—suggesting a pragmatic rather than ideological approach to the issue.

3 Maxmind's basic country-level service costs US$0.0001 per query—a modest but substantive cost for websites attracting millions of hits daily.

4 The tech blog Techdirt recently did an accuracy test of various city-level geolocation services (which produced rather mixed results), concluding that "these tools are as accurate as taking a dart and throwing it not at a map on the wall, but at a Google map display on your computer screen" (Norton 2016).

5 Once extremely popular, these add-ons fell out of favor somewhat after it was revealed in 2015 that Hola's parent company had been selling access to users' bandwidth for botnet attacks. ISP-level unblockers are another option, though more obscure than consumer-facing devices. The Auckland company Bypass developed a tool called Global Mode that could circumvent geoblocking for customers of participating ISPs. Global Mode was trialed by a few ISPs in Singapore, New Zealand, and Australia, then discontinued in 2015 following pressure from rights-holders (Pullar-Strecker 2015).

6 The content of these emails has been previously reported in the trade press. The full text of the emails was subsequently archived by WikiLeaks in April 2015 and can be accessed on the WikiLeaks Sony Archives website.

7 Various other references to this issue can also be found in emails dated throughout 2013 and 2014, in which very senior people at

Sony voiced similar concerns about the Netflix geofiltering issue. Clearly the issue had been simmering internally within Sony for some time before becoming public in 2014.

8 See, for example, the European Union's Digital Single Market policy and the Australian House of Representatives Standing Committee on Infrastructure and Communications report *At What Cost? IT Pricing and the "Australia Tax"* (2013).

CONCLUSIONS

1 Live streamers monetize their channels by receiving virtual "gifts" from their fans.

BIBLIOGRAPHY

Abbate, Janet. 1999. *Inventing the Internet*. Cambridge, MA: MIT Press.

Aglionby, John, and Matthew Garrahan. 2016. "Kenya Threatens to Ban Netflix over 'Inappropriate Content.'" *Financial Times*, January 21. www.ft.com/content/9e97edf0-bf71-11e5-846f-79b0e3d20eaf.

Alexander, Neta. 2016. "Catered to Your Future Self: Netflix's 'Predictive Personalization' and the Mathematization of Taste." In *The Netflix Effect: Technology and Entertainment in the Twenty-First Century*, edited by Kevin McDonald and Daniel Smith-Rowsey, 81–97. New York: Bloomsbury Academic.

Amatriain, Xavier, and Justin Basilico. 2013. "System Architectures for Personalization and Recommendation." *Netflix Tech Blog*, March 27. techblog.netflix.com/2013/03/system-architectures-for.html.

Anderson, John. 2016. *An Over-the-Top Exemption: It's Time to Fairly Tax and Regulate the New Internet Media Services*. Report. Ottawa: Canadian Centre for Policy Alternatives.

Associated Press. 2004. "Geolocation Tech Slices, Dices World Wide Web." *Augusta Chronicle*, July 12. chronicle.augusta.com/stories/2004/07/12/liv_421904.shtml.

Australian House of Representatives Standing Committee on Infrastructure and Communications. 2013. *At What Cost? IT Pricing and the "Australia Tax."* Report.

Avari, Jamshed. 2016. "Netflix Is Here, and No, It Is Not Censoring Any Content." *Gadets360*, January 7. gadgets.ndtv.com/tv/features/netflix-is-here-and-no-it-is-not-censoring-any-content-786669.

Banks, Jack. 1997. *Monopoly Television: MTV's Quest to Control the Music*. Boulder, CO: Westview Press.

Barker, Chris. 1997. *Global Television: An Introduction*. Malden, MA: Blackwell.

Beer, David. 2013. *Popular Culture and New Media: The Politics of Circulation*. New York: Palgrave Macmillan.

Beirne, Rebecca. 2015. "Piracy, Geoblocking, and Australian Access to Niche Independent Cinema." *Popular Communication* 13 (1): 18–31.

Bennett, James. 2011. "Introduction: Television as Digital Media." In *Television as Digital Media*, edited by James Bennett and Nikki Strange, 1–27. Durham, NC: Duke University Press.

Bennett, James, and Nikki Strange, eds. 2011. *Television as Digital Media*. Durham, NC: Duke University Press.

Blakley, Johanna. 2016. "Technologies of Taste." *IEEE Technology and Society Magazine*, December, 39–43.

Blaney, Martin. 2013. "Netflix 'Open' to Euro Investment." *Screen International*, November 22. www.screendaily.com/news/netflix -open-to-euro-investment/5063919.article.

Block, Alex Ben. 2012. "Netflix's Ted Sarandos Explains Original Content Strategy." *Hollywood Reporter*, April 7. www.hollywoodre porter.com/news/netflix-ted-sarandos-original-content-309275.

Blum, Andrew. 2012. *Tubes: A Journey to the Center of the Internet*. New York: Ecco.

Boddy, William. 2004. "Interactive Television and Advertising Form in Contemporary U.S. Television." In *Television after TV: Essays on a Medium in Transition*, edited by Lynn Spigel and Jan Olsson, 113–132. Durham, NC: Duke University Press.

Böttger, Timm, Felix Cuadrado, Gareth Tyson, Ignacio Castro, and Steve Uhlig. 2016. "Open Connect Everywhere: A Glimpse at the Internet Ecosystem through the Lens of the Netflix CDN." arXiv Working Paper. eecs.qmul.ac.uk/~boettget/mapping-netflix -coseners16.pdf.

Bogost, Ian, and Nick Montfort. 2009a. "New Media as Material Constraint: An Introduction to Platform Studies." Paper read at the 1st International HASTAC Conference, at Duke University, Durham, NC. bogost.com/downloads/Bogost%20Montfort%20 HASTAC.pdf.

———. 2009b. "Platform Studies: Frequently Questioned Answers." In *Proceedings of the Digital Arts and Culture Conference 2009*, University of California-Irvine, December 14, 2009. pdf.textfiles. com/academics/bogost_montfort_dac_2009.pdf.

Borenstein, Eliot. 2016. "No Netflix, No Chill: In Russia, It's Better to Purge than to Binge." *Huffington Post*, n.d. www.huffingtonpost .com/entry/no-netflix-no-chill-in-ru_b_10636080.html.

Boyd, Danah, and Kate Crawford. 2012. "Critical Questions for Big Data." *Information, Communication and Society* 15 (5): 662–679.

Braun, Joshua. 2013. "Going Over the Top: Online Television Distribution as Sociotechnical System." *Communication, Culture and Critique* 6 (3): 432–458.

———. 2015. *This Program Is Brought to You by . . . Distributing Television News Online*. New Haven, CT: Yale University Press.

Brodkin, Jon. 2016. "Netflix Asks FCC to Declare Data Caps 'Unreasonable.'" *Ars Technica*, September 13. arstechnica.com /information-technology/2016/09/netflix-asks-fcc-to-declare -data-caps-unreasonable/.

Brooker, Will. 2001. "Living on *Dawson's Creek*: Teen Viewers, Cultural Convergence, and Television Overflow." *International Journal of Cultural Studies* 4 (4): 456–472.

Buonanno, Milly. 2007. *The Age of Television: Experiences and Theories*. Bristol: Intellect.

Burgess, Jean, and Nancy Baym. 2016. "@RT#: Towards a Platform Biography of Twitter." In *Platform Studies: The Rules of Engagement*, 9–11. Selected papers of AoIR 2016: The 17th Annual Conference of the Association of Internet Researchers, Berlin, October 5–8, 2016.

Burgess, Jean, and Joshua Green. 2009. *YouTube: Online Video and Participatory Culture*. Cambridge: Polity Press.

Burrington, Ingrid. 2016. "Why Amazon's Data Centers Are Hidden in Spy Country." *The Atlantic*, January 8. www.theatlantic.com /technology/archive/2016/01/amazon-web-services-data-center /423147/.

Burroughs, Benjamin. 2015. "Streaming Media: Audience and Industry Shifts in a Networked Society." PhD thesis, University of Iowa.Cain, Rob. 2015. "Netflix Japan Launch Over-Hyped, Over-Valued." August 7, www.forbes.com/sites/robcain/2015/08/07 /netflix-japan-launch-over-hyped-over-valued.

Castells, Manuel. 1996. *The Rise of the Network Society*. Cambridge: Blackwell.

———. 2001. *The Internet Galaxy: Reflections on the Internet, Business, and Society*. Oxford: Oxford University Press.

Chalaby, Jean K. 2005. *Transnational Television Worldwide: Towards a New Media Order*. London: I. B. Tauris.

———. 2009. *Transnational Television in Europe: Reconfiguring Global Communications Networks*. London: I. B. Tauris.

———. 2016. *The Format Age: Television's Entertainment Revolution*. Cambridge: Polity Press.

Christian, Aymar Jean. 2017. *Open TV: Innovation Beyond Hollywood and the Rise of Web Television*. New York: New York University Press.

Cicconi, Jim. 2014. "Who Should Pay for Netflix?" *AT&T Public Policy Blog*, March 21. www.attpublicpolicy.com/consumers-2/who -should-pay-for-netflix/.

Clark, Jessica, Nick Couldry, Abigail T. De Kosnik, Tarleton Gillespie, Henry Jenkins, Christopher Kelty, Zizi Papacharissi, Alison Powell, and José van Dijck. 2014. "Participations: Dialogues on the Participatory Promise of Contemporary Culture and Politics | Part 5: Platforms." *International Journal of Communication* 8:1446–1473.

Collins, Richard. 1993. *Audiovisual and Broadcasting Policy in the European Community*. London: University of North London Press.

———. 1998. *From Satellite to Single Market: New Communication Technology and European Public Service Television*. New York: Routledge.

Cunningham, Stuart, and David Craig. 2019. *Social Media Entertainment: The New Industry at the Intersection of Hollywood and Silicon Valley*. New York: New York University Press.

Cunningham, Stuart, and Jon Silver. 2013. *Screen Distribution and the New King Kongs of the Online World*. Basingstoke: Palgrave.

Curtin, Michael. 2007. *Playing to the World's Biggest Audience: The Globalization of Chinese Film and TV*. Berkeley: University of California Press.

Davies, Lyell. 2016. "Netflix and the Coalition for an Open Internet." In *The Netflix Effect: Technology and Entertainment in the Twenty-First Century*, edited by Kevin McDonald and Daniel Smith-Rowsey, 15–31. New York: Bloomsbury Academic.

de Sola Pool, Ithiel. 1990. *Technologies without Boundaries: On Telecommunications in a Global Age*. Cambridge, MA: Belknap Press.

de Valck, Marijke, and Jan Teurlings, eds. 2013. *After the Break: Television Theory Today*. Amsterdam: Amsterdam University Press.

DiMaggio, Paul, and Eszter Hargittai. 2001. "From the 'Digital Divide' to 'Digital Inequality': Studying Internet Use as Penetration

Increases." Working Paper 15, Center for Arts and Cultural Policy Studies, Princeton University.

Donoghue, Courtney Brannon. 2017. *Localizing Hollywood*. London: Palgrave/British Film Institute.

The Economist. 2017. "How to Devise the Perfect Recommendation Algorithm." February 9. www.economist.com/news/special -report/21716464-recommendations-must-be-neither-too-famil iar-nor-too-novel-how-devise-perfect.

Edwards, Paul, Steven J. Jackson, Geoffrey C. Bowker, and Cory Knobel. 2007. "Understanding Infrastructure: Dynamics, Tensions, and Design." Report of the Workshop "History and Theory of Infrastructure: Lessons for New Scientific Cyberinfrastructures," University of Michigan. hdl.handle.net/2027.42/49353.

Edwards, Paul N. 2003. "Infrastructure and Modernity: Force, Time, and Social Organization in the History of Sociotechnical Systems." In *Modernity and Technology*, edited by Thomas J. Misa, Philip Brey, and Andrew Feenberg, 185–225. Cambridge, MA: MIT Press.

Elkins, Evan. 2018. "Powered by Netflix: Speed-Test Services and Video-on-Demand's Global Development Projects." *Media, Culture and Society,* pre-published online February 1, doi .org/10.1177/0163443718754649.

Elsaesser, Thomas, and Malte Hagener. 2015. *Film Theory: An Introduction through the Senses.* 2nd ed. London: Routledge.

Ernesto. 2017a. "Netflix VPN Crackdown, a Year of Frustrations." *Torrent Freak*, January 21. torrentfreak.com/netflix-vpn-crackdown-a -year-of-frustrations-170120/.

———. 2017b. "Netflix Gets Serious about Its Anti-piracy Efforts." *Torrent Freak*, March 24. torrentfreak.com/netflix-gets-serious -about-its-anti-piracy-efforts-170324/.

———. 2018. "Netflix, Amazon and Hollywood Sue Kodi-Powered Dragon Box over Piracy." *Torrent Freak*, January 11. torrentfreak .com/netflix-amazon-and-hollywood-sue-kodi-powered-dragon -box-over-piracy-180111/.

European Commission. 2014. "Promotion of European Works in Practice." Briefing paper. ec.europa.eu/newsroom/dae/document .cfm?action=display&doc_id=11459.

Farr, Brittany. 2016. "Seeing Blackness in Prison: Understanding Prison Diversity on Netflix's *Orange Is the New Black*." In *The*

Netflix Effect: Technology and Entertainment in the Twenty-First Century, edited by Kevin McDonald and Daniel Smith-Rowsey, 155–170. New York: Bloomsbury Academic.

Fetner, Chris, and Danny Sheehan. 2017. "The Netflix HERMES Test: Quality Subtitling at Scale." *Netflix Tech Blog*, March 31. medium .com/netflix-techblog/the-netflix-hermes-test-quality-subtitling-at -scale-dccea2682aef.

Flew, Terry. 2007. *Understanding Global Media*. New York: Palgrave.

Flew, Terry, Petros Iosifidis, and Jeanette Steemers, eds. 2016. *Global Media and National Policies*. Basingstoke: Palgrave Macmillan.

Flint, Joe, and Shalini Ramachandran. 2017. "Netflix: The Monster That's Eating Hollywood." *Wall Street Journal*, March 24. www.wsj.com/ articles/netflix-the-monster-thats-eating-hollywood-1490370059.

Florance, Ken. 2016. "How Netflix Works with ISPs Around the Globe to Deliver a Great Viewing Experience." Netflix press release, March 17. media.netflix.com/en/company-blog/how-net flix-works-with-isps-around-the-globe-to-deliver-a-great-viewing -experience.

Fontaine, Gilles, and Christian Grece. 2016. *Origin of Films and TV Content in VOD Catalogues in the EU and Visibility of Films on VOD Services*. Strasbourg: European Audiovisual Observatory.

Frater, Patrick. 2016. "Netflix Faces Challenges as It Plans a Global Launch, Particularly in Asia." *Variety*, February 5. variety.com /2016/digital/global/netflix-asia-challenges-global-launch-120169 6252/.

Gerlitz, Carolin, and Anne Helmond. 2013. "The Like Economy: So-cial Buttons and the Data-Intensive Web." *New Media and Society* 15 (8): 1348–1365.

Gillespie, Marie. 1995. *Television, Ethnicity and Cultural Change*. London: Routledge.

Gillespie, Tarleton. 2010. "The Politics of 'Platforms.'" *New Media and Society* 12 (3): 347–364.

———. 2017. "Regulation of and by Platforms." In *Sage Handbook of Social Media*, edited by Jean Burgess, Thomas Poell, and Alice Marwick, 254–278. Thousand Oaks, CA: Sage.

Global Web Index. 2014. "Report on VPN Usage—The Missing Billion." insight.globalwebindex.net/hs-fs/hub/304927/file -1631708567-pdf.

Globerman, Steven. 2014. *Canadian Content Is Dead; Long Live Canadian Content!* Report. Calgary: Fraser Institute.

Goldsmith, Jack, and Tim Wu. 2006. *Who Controls the Internet? Illusions of a Borderless World*. New York: Oxford University Press.

Golumbia, David. 2016. *The Politics of Bitcoin*. Minneapolis: University of Minnesota Press.

Goodfellow, Melanie. 2015. "How is Netflix performing after a year in France?" Screen Daily, September 16, www.screendaily.com/news /how-is-netflix-performing-after-a-year-in-france/5093186.article.

Google. 2017. "Transparency Report." transparencyreport.google.com/.

Government of France. 2015. "Consultation sur la directive 2010/13/ UE relative aux services de médias audiovisuels (directive SMA); Réponse de la France." (n.d.), ec.europa.eu/newsroom/dae/docu ment.cfm?action=display&doc_id=11850.

Graham, Stephen, and Simon Marvin. 1996. *Telecommunications and the City: Electronic Spaces, Urban Places*. London: Routledge.

Gray, Jonathan. 2009. *Television Entertainment*. Hoboken, NJ: Routledge.

Gunawan S., Arif. 2016. "Indonesia's Telkom Effectively Blocks Netflix." *Jakarta Post*, January 27. www.thejakartapost.com/news/2016/01/27 /indonesia-s-telkom-effectively-blocks-netflix.html.

Gunther, Marc. 2004. "MTV's Passage to India." *Fortune*, August 9, 116.

Hadfield, James. 2015. "Tokyo: Streaming Video Still Flowing Slowly in Japan." Variety, October 20, www.variety.com/2015/digital/asia /streaming-video-still-flowing-slowly-in-japan-1201621949/.

Hanke, Robert. 1998. "'Yo quiero mi MTV!': Making Music Television for Latin America." In *Mapping the Beat: Popular Music and Contemporary Theory*, edited by Thomas Swiss, John Sloop, and Andrew Herman, 219–246. Malden, MA: Blackwell.

Hastings, Reed. 2014. "How to Save the Net: Don't Give in to Big ISPs." *Wired*, August 19. www.wired.com/2014/08/save-the-net-reed -hastings/.

Hilmes, Michele. 2012. *Network Nations: A Transnational History of British and American Broadcasting*. New York: Routledge.

Holt, Jennifer, and Patrick Vonderau. 2015. "'Where the Internet Lives': Data Centres as Cloud Infrastructure." In *Signal Traffic: Critical Studies of Media Infrastructures*, edited by Lisa Parks and Nicole Starosielski, 71–93. Urbana: University of Illinois Press.

Innis, Harold. 1951. *The Bias of Communication*. Toronto: University of Toronto Press.

Iwabuchi, Koichi, ed. 2004. *Feeling Asian Modernities: Transnational Consumption of Japanese TV Dramas*. Hong Kong: Hong Kong University Press.

Jackson, Steven J. 2013. "Rethinking Repair." In *Media Technologies: Essays on Communication, Materiality and Society*, edited by Tarleton Gillespie, Pablo J. Boczkowski, and Kirsten A. Foot, 221–239. Cambridge, MA: MIT Press.

Jackson, Steven J., Paul Edwards, Geoffrey Bowker, and Cory P. Knobe. 2007. "Understanding Infrastructure: History, Heuristics and Cyberinfrastructure Policy." *First Monday* 12 (6). firstmonday.org/ojs/index.php/fm/article/view/1904.

Jakubowicz, Karol. 1994. "Internationalization of Television in Central and Eastern Europe." In *Central and Eastern Europe: Audiovisual Landscape and Copyright Legislation*, edited by Karol Jakubowicz and Pierre Jeanray. 13–20. Apeldoorn: Maklo Uitgevers Antwerpen.

Jarnes, Mark. 2016. "Can Netflix's 'Hibana' Spark a Revolution in Japanese TV?" *Japan Times*, June 9. www.japantimes.co.jp/culture/2016/06/09/tv/can-netflixs-hibana-spark-revolution-japanese-tv/.

Jentreau, Katell. 2017. "Localizing for the Netflix Effect." *Globally Speaking Radio* podcast, June 21. www.globallyspeakingradio.com/podcast/podcast-030-localizing-for-the-netflix-effect.

Johns, Adrian. 2011. *Death of a Pirate: British Radio and the Making of the Information Age*. New York: W. W. Norton.

Johnson, Catherine. 2007. "Telebranding in TVIII: The Network as Brand and the Programme as Brand." *New Review of Film and Television Studies* 5 (1): 5–24.

———. 2017. "Defining Television in an Online Video Ecosystem." Paper presented at Trans TV conference, University of Westminster, London, September 15.

Katz, Elihu, and George Wedell. 1977. *Broadcasting in the Third World: Promise and Performance*. Cambridge, MA: Harvard University Press.

Keating, Gina. 2012. *Netflixed: The Epic Battle for America's Eyeballs*. New York: Portfolio/Penguin.

Kelsey, John. 2010. "The Year in Television." *Artforum International* 49 (4): 230–233.

Keslassy, Elsa. 2013. "Netflix's French Resistance: The Market Reportedly Ripe for the Streaming Service's Next Invasion Boasts Huge Barriers to Entry." *Variety*, November 26, 28.

Kitchin, Rob, and Martin Dodge. 2011. *Code/Space: Software and Everyday Life*. Cambridge, MA: MIT Press.

Kozlov, Vladimir. 2016. "Netflix May Have to Suspend Operations in Russia, Says Government Minister." *Hollywood Reporter*, February 10. www.hollywoodreporter.com/news/netflix-may-have-suspend-operations-864130.

———. 2017. "Why Russia's Foreign Ownership Restrictions on Streamers Do Not Affect Netflix." *Hollywood Reporter*, August 31. www.hollywoodreporter.com/news/netflix-exempt-russias-foreign-ownership-restrictions-1034219.

Kraidy, Marwan M. 2008. "Reality TV and Multiple Arab Modernities: A Theoretical Exploration." *Middle East Journal of Culture and Communication* 1:49–59.

Kuo, Lily. 2016. "Kenya's Film Regulator Is Calling Netflix a Threat to the Country's National Security." *Quartz*, January 20. qz.com/598521/kenyas-film-regulator-is-calling-netflix-a-threat-to-the-countrys-national-security/.

Lampland, Martha, and Susan Leigh Star, eds. 2008. *Standards and Their Stories: How Quantifying, Classifying, and Formalizing Practices Shape Everyday Life*. Ithaca, NY: Cornell University Press.

Larkin, Brian. 2008. *Signal and Noise: Media, Infrastructure, and Urban Culture in Nigeria*. Durham, NC: Duke University Press.

———. 2013. "The Politics and Poetics of Infrastructure." *Annual Review of Anthropology* 42:327–343.

Larsen, Peter. 1990. *Import/Export: International Flow of Television Fiction*. Paris: UNESCO.

Lash, Scott, and Celia Lury. 2007. *Global Culture Industry: The Mediation of Things*. Cambridge: Polity Press.

Leaver, Tama. 2008. "Watching *Battlestar Galactica* in Australia and the Tyranny of Digital Distance." *Media International Australia* 126 (1): 145–154.

Leonard, Andrew. 2013. "How Netflix Is Turning Viewers into Puppets." *Salon*, February 2. www.salon.com/2013/02/01/how_netflix_is _turning_viewers_into_puppets/.

Lev-Ram, Michal. 2016. "The Problem with Silicon Valley? Unbridled Power." *Fortune*, March 9. fortune.com/2016/03/09/john-land-graf-ceo-fx/.

Liebes, Tamar, and Elihu Katz. 1990. *The Export of Meaning: Cross-Cultural Readings of Dallas*. London: Oxford University Press.

Lobato, Ramon, and James Meese. 2016. *Geoblocking and Global Video Culture*. Amsterdam: Institute of Network Cultures.

Lobato, Ramon, and Alexa Scarlata. 2017. "Australian content in SVOD catalogs: availability and discoverability. Submission to the Australian and Children's Screen Content Review." September, apo .org.au/node/134926.

Lotz, Amanda. 2014. *The Television Will Be Revolutionized*. 2nd ed. New York: New York University Press.

———. 2017a. *Portals: A Treatise on Internet-Distributed Television*. Ann Arbor: University of Michigan Press.

———. 2017b. *We Now Disrupt This Broadcast: How Cable Transformed Television and the Internet Revolutionized It All*. Cambridge, MA: MIT Press.

Madappa, Shashi, Vu Nguyen, Scott Mansfield, Sridhar Enugula, Allan Pratt, and Faisal Zakaria Siddiqi. 2016. "Caching for a Global Netflix." Netflix Tech Blog, March 1. medium.com/netflix-techblog/ caching-for-a-global-netflix-7bcc457012f1.

Madrigal, Alexis C. 2014. "How Netflix Reverse Engineered Hollywood." *The Atlantic*, January 2. www.theatlantic.com/technology /archive/2014/01/how-netflix-reverse-engineered-hollywood/282679/.

Mattelart, Armand. 1979. *Multinational Corporations and the Control of Culture: The Ideological Apparatuses of Imperialism*. Brighton: Harvester Press.

McEntree, Kevin. 2010. "Why We Use and Contribute to Open Source Software." *Netflix Tech Blog*, December 10. techblog.netflix .com/2010/12/why-we-use-and-contribute-to-open.html.

Mendes Moreira de Sa, Vanessa. 2016. "Brazil: Netflix, VPNs and the 'Paying' Pirates." In *Geoblocking and Global Video Culture*, edited by Ramon Lobato and James Meese, 158–167. Amsterdam: Institute of Network Cultures.

Miller, Daniel. 1992. "The Young and the Restless in Trinidad: A Case of the Local and the Global in Mass Consumption." In *Consuming Technologies: Media and Information in Domestic Spaces*, edited by Roger Silverstone and Eric Hirsch, 163–182. London: Routledge.

Miller, Peter H., and Randal Rudniski. 2012. *Market Impact and Indicators of Over the Top Television in Canada: 2012*. Report for Canadian Radio-Television and Telecommunications Commission. www.crtc.gc.ca/eng/publications/reports/rp120330.htm.

Miller, Toby. 2010. *Television Studies: The Basics*. London: Routledge.

Miller, Toby, Nitin Govil, John McMurria, Richard Maxwell, and Ting Wang. 2005. *Global Hollywood 2*. London: British Film Institute.

Montfort, Nick, and Ian Bogost. 2009. *Racing the Beam: The Atari Video Computer System*. Cambridge, MA: MIT Press.

Moran, Albert. 1998. *Copycat TV: Globalisation, Program Formats and Cultural Identity*. Luton: University of Luton Press.

———. 2009. *New Flows in Global TV*. Bristol: Intellect.

Morley, David, and Kevin Robins. 1995. *Space of Identity: Global Media, Electronic Landscapes, and Cultural Boundaries*. New York: Routledge.

Mowlana, Hamid. 1985. *International Flow of Information: A Global Report and Analysis*. Paris: UNESCO.

Murphy, Sheila C. 2011. *How Television Invented New Media*. New Brunswick, NJ: Rutgers University Press.

Mytton, Graham, Ruth Teer-Tomaselli, and André-Jean Tudesq. 2005. "Transnational Television in Sub-Saharan Africa." In *Transnational Television Worldwide: Towards a New Media Order*, edited by Jean Chalaby, 96–127. London: I. B. Tauris.

Nair, Mayukh. 2017. "How Netflix Works: The (Hugely Simplified) Complex Stuff That Happens Every Time You Hit Play." *Refraction: Tech and Everything*, blog post, October 17. medium.com/refraction-tech-everything/how-netflix-works-the-hugely-simplified-complex-stuff-that-happens-every-time-you-hit-play-3a40c9be254b.

Negroponte, Nicholas. 1995. *Being Digital*. Rydalmere: Hodder and Stoughton.

Netflix. 2014. "Transcript of NFLX Q3 2014 Netflix Inc Earnings Call." ir.netflix.com/results.cfm.

———. 2015a. "Transcript of NFLX Q1 2015 Netflix Inc Earnings Call." ir.netflix.com/results.cfm.

———. 2015b. "Netflix Is Available in Cuba." Press release, February 9. media.netflix.com/en/press-releases/netflix-is-available-in-cuba.

———. 2015c. "Consultation on Directive 2010/13/EU on Audiovisual Media Services (AVMSD)." Submission by Netflix International B.V., (n.d.), http://ec.europa.eu/newsroom/dae/document .cfm?action=display&doc_id=11459.

———. 2016. "Transcript of NFLX Q1 2015 Netflix Inc Earnings Call." ir.netflix.com/results.cfm.

———. 2017a. "Netflix Is Looking for the Best Translators Around the Globe." Press release, March 30. media.netflix.com/en/company -blog/netflix-is-looking-for-the-best-translators-around-the-globe.

———. 2017b. "What Netflix's Half a Billion CAD Investment in Canada Is Really About." Press release, October 10. https://media. netflix.com/en/company-blog/what-netflixs-half-a-billion-cad -investment-in-canada-is-really-about.

———. 2017c. "Top Investor Questions." Netflix Investor Relations website, https://ir.netflix.com/top-investor-questions.

Nordenstreng, Kaarle, and Herbert I. Schiller, eds. 1979. *National Sovereignty and International Communication*. Norwood, NJ: Ablex.

Nordenstreng, Kaarle, and Tapio Varis. 1974. *Television Traffic—A One-Way Street? A Survey and Analysis of the International Flow of Television Programme Material*. Paris: UNESCO Reports and Papers on Mass Communication.

Norton, Andrew. 2016. "How Bad Are Geolocation Tools? Really, Really Bad." *TechDirt*, April 15. www.techdirt.com/articles /20160413/12012834171/how-bad-are-geolocation-tools-really -really-bad.shtml.

O'Brien, Chris. 2014. "Netflix Struggles to Win Over Skeptics in Film-Loving France." *Los Angeles Times*, September 28. www.latimes .com/business/la-fi-ct-netflix-france-20140928-story.html.

Palacin, Manuel, Miquel Oliver, Jorge Infante, Simon Oechsner and Alex Bikfalvi. 2013. "The Impact of Content Delivery Networks on the Internet Ecosystem." *Journal of Information Policy* 3:304–330.

Pariser, Eli. 2011. *The Filter Bubble: What the Internet Is Hiding from You*. New York: Penguin Press.

Parks, Lisa. 2004. "Flexible Microcasting: Gender, Generation, and Television-Internet Convergence." In *Television after TV: Essays on a Medium in Transition*, edited by Lynn Spigel and Jan Olsson, 133–156. Durham, NC: Duke University Press.

———. 2005. *Cultures in Orbit: Satellites and the Televisual*. Durham, NC: Duke University Press.

Parks, Lisa, and Shanti Kumar, eds. 2003. *Planet TV: A Global Television Reader*. New York: New York University Press.

Parks, Lisa, and Nicole Starosielski. 2015. *Signal Traffic: Critical Studies of Media Infrastructures*. Urbana: University of Illinois Press.

Pearson, Roberta. 2011. "Cult Television as Digital Television's Cutting Edge." In *Television as Digital Media*, edited by James Bennett and Nikki Strange, 105–131. Durham, NC: Duke University Press.

Pedwell, Terry. 2014. "CRTC to Netflix: Since You Won't Co-operate, We'll Ignore You." CBCNews, September 29. www.cbc.ca/news /business/crtc-to-netflix-since-you-won-t-co-operate-we-ll-ignore -you-1.2781748.

Pertierra, Anna Cristina, and Graeme Turner. 2013. *Locating Television: Zones of Consumption*. New York: Routledge.

Ploman, Edward W. 1979. "Satellite Broadcasting, National Sovereignty, and Free Flow of Information." In *National Sovereignty and International Communication*, edited by Kaarle Nordenstreng and Herbert I. Schiller, 154–165. Norwood, NJ: Ablex.

Pullar-Strecker, Tom. 2015. "TV Companies Go to Court over Global Mode." *Stuff.co.nz*, April 20. www.stuff.co.nz/business/industries /67879996/TV-companies-go-to-court-over-Global-Mode.

Rajab, Ramadhan. 2016. "Netflix Content Immoral, Says Films Board." *The Star* (Kenya), January 21. www.the-star.co.ke/news/2016/01/21 /netflix-content-immoral-says-films-board_c1280034.

Rao, Vyjayanthi V. 2014. "Infra-city: Speculations on Flux and History in Infrastructure-Making." In *Infrastructural Lives: Urban Infrastructure in Context*, edited by Stephen Graham and Colin McFarlane, 39–58. London: Routledge.

Rayburn, Dan. 2014. "Chart Shows Which Content Owners Have Direct Interconnect Deals with ISPs." *Streamingmedia.com* blog, May 21. blog.streamingmedia.com/2014/05/chart-shows-which -content-owners-have-direct-interconnect-deals-with-isps.html.

Reilly, Claire. 2014. "Rights Holders Move to Block Australian Access to US Netflix." *CNET*, September 17. www.cnet.com/au/news /rights-holders-move-to-block-us-netflix-viewing-in-australia/.

Robertson, Susan Krashinsky. 2017. "Netflix Leads Streaming Services in Canada." *The Globe and Mail*, October 20. www.theglobe andmail.com/report-on-business/industry-news/marketing /netflix-leads-streaming-services-in-canada/article36678928/.

Rodríguez, Fidel. 2016. "Cuba: *Videos to the Left*—Circumvention Practices and Audiovisual Ecologies." In *Geoblocking and Global Video Culture*, edited by Ramon Lobato and James Meese, 178–189. Amsterdam: Institute of Network Cultures.

Roettgers, Janko. 2015. "Inside Netflix's Plan to Boost Streaming Quality and Unclog the Internet (Exclusive)." *Variety*, December 14. variety.com/2015/digital/news/netflix-better-streaming-quality -1201661116/.

———. 2017. "How Netflix Wants to Rule the World: A Behind-the-Scenes Look at a Global TV Network." *Variety*, March 18. variety .com/2017/digital/news/netflix-lab-day-behind-the-scenes -1202011105/.

Rossiter, Ned. 2016. *Software, Infrastructure, Labor: A Media Theory of Logistical Nightmares*. London: Routledge.

Roy Morgan Research. 2016. "More Australians now have SVOD than Foxtel." Press release (finding no. 6957), September 8. www .roymorgan.com/findings/6957-svod-overtakes-foxtel-pay-tv-in -australia-august-2016-201609081005.

Roy Morgan Research. 2017. "Over 1 in 3 Australians Now Have Netflix as Subscriptions Jump 20 Percent in First Quarter of 2017." Press release (finding no. 7242), June 8. www.roymorgan.com /findings/7242-netflix-subscriptions-march-2017-201706080957.

Sakr, Naomi. 2001. *Satellite Realms: Transnational Television, Globalization and the Middle East*. London: I. B. Tauris.

Sandvig, Christian. 2015. "The Internet as Infrastructure." In *The Oxford Handbook of Internet Studies*, edited by William H. Dutton, 86–106. Oxford: Oxford University Press.

Sandvine. 2016a. *Global Internet Phenomena 2016: Latin America and North America*. Waterloo: Sandvine Incorporated.

———. 2016b. *Global Internet Phenomena 2016: Africa, Asia-Pacific, and the Middle East*. Waterloo: Sandvine Incorporated.

Santana. Kenny. 2003. "MTV Goes to Asia." Global Policy Forum, August 12, www.globalpolicy.org/component/content/article /162/27623.html.

Sarkari, John. 2016. "Netflix Is for Premium Customers in India, Says Company." *Times of India*, February 2. timesofindia.indiatimes. com/business/india-business/Netflix-is-for-premium-customers -in-India-says-company/articleshow/50813180.cms.

Scarlata, Alexa. 2015. "Australian Streaming Services and the Relationship between Viewing Data and Local Television Drama Production." Paper presented at Australia Screen Production Education and Research Association Conference, Flinders University, Adelaide, Australia, 2015.

Schellevis, Door Joost. 2013. "Netflix baseert aanbod deels op populariteit video's op piraterijsites." *Tweakers.net*, September 14. tweakers.net/nieuws/91282/netflix-baseert-aanbod-deels-op-popu lariteit-videos-op-piraterijsites.html.

Schiller, Herbert. 1969. *Mass Communications and American Empire*. Boulder, CO: Westview Press.

Shacklock, Zoë. 2016. "On (Not) Watching Outlander in the United Kingdom." *Visual Culture in Britain* 17 (3): 311–328.

Shaw, Lucas. 2015. "Can Netflix Become Must-See TV in Japan?" August 27, https://www.bloomberg.com/news/articles/2015-08-27 /can-netflix-become-must-see-tv-in-japan-.Sinclair, John, Liz Jacka, and Stuart Cunningham, eds. 1995. *New Patterns in Global Television: Peripheral Vision*. New York: Oxford University Press.

Solsman, Joan E. 2017. "Netflix Is Hijacking 1 Billion Hours of Our Lives Each Week." *CNet*, April 17. www.cnet.com/news/netflix -billion-hours-a-week-adam-sandler/.

Spigel, Lynn. 2005. "TV's Next Season?" *Cinema Journal* 45 (1): 83–90.

Spigel, Lynn, and Jan Olsson. 2004. *Television after TV: Essays on a Medium in Transition*. Durham, NC: Duke University Press.

Star, Susan Leigh, and Geoffrey C. Bowker. 2000. *Sorting Things Out: Classification and Its Consequences*. Cambridge, MA: MIT Press.

Star, Susan Leigh, and Karen Ruhleder. 1996. "Steps toward an Ecology of Infrastructure: Design and Access for Large Information Spaces." *Information Systems Research* 7 (1): 111–134.

Starosielski, Nicole. 2015. *The Undersea Network*. Durham, NC: Duke University Press.

Sterne, Jonathan. 2012. *MP3: The Meaning of a Format.* Durham, NC: Duke University Press.

Stewart, Mark. 2016. "The Myth of Televisual Ubiquity." *Television and New Media* 17 (8): 691–705.

Straubhaar, Joseph D. 1991. "Beyond Media Imperialism: Asymmetrical Interdependence and Cultural Proximity." *Critical Studies in Mass Communication* 8 (1): 39–59.

———. 2007. *World Television: From Global to Local.* Los Angeles: Sage.

Svantesson, Dan Jerker B. 2004. "Geo-location Technologies and Other Means of Placing Borders on the 'Borderless' Internet." *Journal of Computer and Information Law* 23 (1): 101–140.

Taylor, Kate. 2016. "Melanie Joly's Fight for Canadian Culture Is Neither Easy nor Popular." *The Globe and Mail*, December 16. www .theglobeandmail.com/arts/art-and-architecture/melanie -jolys-fight-for-canadian-culture-is-neither-easy-nor-popular /article33345113/.

———. 2017. "Mélanie Joly's Netflix Deal Fails to Address the Real Issues for Canadian Content Creators." *The Globe and Mail*, September 28. www.theglobeandmail.com/opinion/melanie-jolys -netflix-deal-fails-to-address-the-real-issues-for-canadian-content -creators/article36428560/.

Thomas, Julian. 2008. "The Old New Television and the New: Digital Transitions at Home." *Media International Australia* 129 (1): 91–103.

Thompson, Kristin. 1985. *Exporting Entertainment: America in the World Film Market, 1907–34.* London: British Film Institute.

Thussu, Daya Kishan, ed. 2006. *Media on the Move: Global Flow and Contra-flow.* London: Routledge.

Todreas, Timothy M. 1999. *Value Creation and Branding in Television's Digital Age.* Westport, CT: Quorum Books.

Tomlinson, John. 1991. *Cultural Imperialism: A Critical Introduction.* London: Pinter.

Tracey, Michael. 1985. "The Poisoned Chalice? International Television and the Idea of Dominance." *Daedalus* 114:17–55.

Trimble, Marketa. 2012. "The Future of Cybertravel: Legal Implications of the Evasion of Geolocation." *Fordham Intellectual Property, Media and Entertainment Law Journal* 22 (3): 567–657.

———. 2016. "Geoblocking, Technical Standards and the Law." In *Geoblocking and Global Video Culture*, edited by Ramon Lobato and James Meese, 54–63. Amsterdam: Institute of Network Cultures.

Tryon, Chuck. 2013. *On-Demand Culture: Digital Delivery and the Future of Movies*. New Brunswick, NJ: Rutgers University Press.

Tunstall, Jeremy. 1977. *The Media Are American: Anglo-American Media in the World*. London: Constable.

Turner, Graeme, and Jinna Tay, eds. 2009. *Television Studies after TV: Understanding Television in the Post-broadcast Era*. London: Routledge.

Ueland, Chris. 2015. "The Stack behind Netflix Scaling." *ScaleScale* blog, November 5. www.scalescale.com/the-stack-behind-netflix -scaling/.

Uricchio, William. 2004. "Television's Next Generation: Technology/ Interface Culture/Flow." In *Television after TV: Essays on a Medium in Transition*, edited by Lynn Spigel and Jan Olsson, 163–182. Durham, NC: Duke University Press.

———. Forthcoming. "Media Specificity and Its Discontents: A Televisual Provocation." In *From Media to Post-media: Continuities and Ruptures*, edited by Nicolas Dulac and André Gaudreault. Paris: Éditions L'Âge d'Homme.

van Dijck, Jose. 2013. *The Culture of Connectivity: A Critical History of Social Media*. Oxford: Oxford University Press.

Wallenstein, Andrew. 2011. "Netflix Shifts toward New World Orders." *Variety*, April 9. variety.com/2011/digital/news/netflix-shifts -toward-new-world-orders-1118035178/.

———. 2013. "How Netflix, HBO May Benefit from Illegal Password-Sharing." *Variety*, June 27. variety.com/2013/biz/news/hbo-netflix-may-benefit-from-illegal-password-sharing-1200502345/.

———. 2014. "Australia's Fevered Streaming Market Isn't Waiting for Netflix." *Variety*, March 19. variety.com/2014/digital/news /australia-flooded-with-streaming-content-providers-and-netflix -isnt-even-there-yet-1201138144/.

Wagman, Ira. 2017. "Talking to Netflix with a Canadian Accent: On Digital Platforms and National Media Policies." In *Reconceptualising Film Policies*, edited by Nolwenn Mingant and Cecilia Tirtaine, 209–221. London: Routledge.

Warf, Barney. 2013. *Global Geographies of the Internet*. Dordrecht: Springer.

Wark, Mackenzie. 1994. *Virtual Geography: Living with Global Media Events*. Bloomington: Indiana University Press.

Wasser, Frederick. 2002. *Veni, Vidi, Video: The Hollywood Empire and the VCR*. Austin: University of Texas Press.

Wells, Allan. 1972. *Picture-Tube Imperialism? The Impact of U.S. Television on Latin America*. Maryknoll, NY: Orbis Books.

Williams, Raymond. 1974. *Television: Technology and Cultural Form*. London: Fontana. Reprinted 2003.

Wolff, Michael. 2015. *Television Is the New Television*. New York: Penguin.

Wu, Tim. 2015. "The Dreaded Bundle Comes to Internet TV." *New Yorker*, May 3. www.newyorker.com/business/currency/the-dreaded-bundle-comes-to-internet-tv.

Zook, Matthew. 2006. "The Geographies of the Internet." *Annual Review of Information Science and Technology* 40 (1): 53–78.

INDEX

ABOUT THE AUTHOR

Ramon Lobato is Senior Research Fellow in Media and Communication at RMIT University, Melbourne. His previous books include *Shadow Economies of Cinema*, *The Informal Media Economy*, and *Geoblocking and Global Video Culture*.

Printed and bound by CPI Group (UK) Ltd, Croydon, CR0 4YY

27/10/2024

14580397-0001